A.V.Yablokov   S.A.Ostroumov

Conservation of Living Nature and Resources:

Problems, Trends, and Prospects

With 36 Figures and 34 Tables

Springer-Verlag
Berlin Heidelberg NewYork
London Paris Tokyo
Hong Kong Barcelona
Budapest

Professor Dr. Alexey V. Yablokov
USSR Supreme Soviet
Committee of Ecology
Moscow, Kremlin, USSR

Dr. Sergey A. Ostroumov
Department of Hydrobiology
Faculty of Biology
Moscow State University
Moscow 119899, USSR

333 · 95

I // c

ISBN 3-540-52096-1 Springer-Verlag Berlin Heidelberg New York
ISBN 0-387-52096-1 Springer-Verlag New York Berlin Heidelberg

© Springer-Verlag Berlin Heidelberg 1991
Printed in Germany

Typesetting: International Typesetters Inc., Manila, Philippines
Printing: Beltz, Hemsbach; Binding: Schäffer, Grünstadt
31/3145-543210 – Printed on acid-free paper

Preface

This book is based on our two books, published in the USSR and translated in a number of other countries *Conservation of living nature: problems and perspectives* (1983) and *The levels of conservation of living nature* (1985). It differs from the vast majority of the numerous books on conservation and environment, which are mainly devoted either to specific problems of conservation of certain taxons, or to problems of conservation of prescribed regions, or to general issues of environmental conservation in toto, while the problems of the conservation of living nature are represented only to a small degree.

Our book is one of the volumes — at present not numerous — that place a high value on the diversity of living nature as the basis for the existence and development of mankind on the Earth. Living nature, besides its own intrinsic value, at all times was, is now and will ever remain the sole, unique and indispensable resource and provider for mankind.

This book is an attempt to analyze the key problems and perspectives of living resource conservation in a non-traditional way. As the main approach we have considered the problems not only according to taxons and types of biomes but also according to the levels of life systems and the factors of man-made influences on nature, and in addition from economical, administrative, managerial and other viewpoints. This many-sided, pluralistic approach reflects the complex, interdepartmental, intertwined character of the many knotty questions of conservation of living nature; any analysis of them based on a more restricted approach is usually inadequately one-sided and shallow.

After an appraisal of the present state, we attempt to look into the near future, to decide how best to be ready to meet this future, from the viewpoint of solving the most urgent conservation problems.

We hope that all readers — from beginner to specialist, from college student to practitioner — will find something interesting and useful in the approach and data that we present here. For this reason the book is addressed not only to biologists, ecologists, conservationists and environmentalists, but also to all who deal with the resources of living nature: those who are involved in agriculture, forestry, fishery, in environmental law and management. The book is also written as a potentially useful source of data and ideas for economists, lawers and decision-makers. The authors also have tried to meet the interests of physicians and those who represent the health sciences. In addition, we think that the book may be of interest to those dealing with conservation issues on the political or international level.

We have used data of vast geographical scope. We have based our analysis on the data closest to us concerning the USSR, and have included also broad information on conservation from Europe and North America, as well as Asia, South America, Africa, Australia and Oceania. While writing the book, we bore in mind our purpose of finding readers in these regions, to whom the book – as we hope – will be interesting and useful for success in conservation work.

During our work we have benefited from advice and material from many colleagues, among them: A.S. Antonov, V.N. Maximov, M.M. Telitchenko (Moscow State University), E.V. Gvozdev (Institute of Zoology, Alma-Ata), the late B.N. Veprintzev (Institute of Biophysics, Puschchino), the late G.G. Vinberg (Zoological Institute, Leningrad), V.N. Bolshakov (Institute of Ecology of Plants and Animals, Sverdlovsk), G.A. Zavarzin (Institute of Microbiology, Moscow), S.G. Priklonsky (Oksky Reserve, Rjazan Region), I.A. Rapoport (Institute of Chemical Physics, Moscow), N.F. Reimers (Institute of Mathematical Economics, Moscow), T.B. Sablina and N.N. Smirnov (Institute of Evolutional Morphology and Ecology, Moscow), L.M. Suschchenja (Institute of Zoology and Parasitology, Minsk), V.E. Flint and M.V. Cherkasova (Institute of Nature Conservation, Moscow), F.R. Shtilmark (Moscow) and N.N. Schcherbak (Institute of Zoology, Kiev). We also thank our colleagues in other countries: W. Klawe (Interamerican Tuna Tropical Commission, USA), C. Stivens (Animal Welfare Institute, USA), J. Barzdo (IUCN Conservation Monitoring Centre, England), H. Hekstra (European Information Centre for Nature Conservation, Strasburg), M. Soule (Michigan State University, USA), D. Simberloff (Florida State University, USA), P. Raven (Missouri Botanical Garden, USA), M. Pavan (Pavia University, Italy), Norman Myers (England), Verne Grant (Texas State University, USA), W.F. Perrin (Southwest Fisheries Research Center, USA), L. Godicl (Maribor University, Yugoslavia), Heinrich Walter (University of Hohenheim, FRG), R.W. Risebourgh (The Bodega Bay Institute, USA), A. Haga (Universitetet i Oslo, Norway), J.P. Raffin (Fédération Française des Sociétés de Protection de la Nature), Y. Fujimaki (Obihiro University, Japan), B. Stugren (Romania).

To all of them we express our gratitude. We acknowledge also the help of Springer-Verlag, who stimulated the creation of this book and made it available for a broad audience of English-speaking readers.

The book is a result of the joint work of both authors. Chapters 3, 4, 6, 7, 10, 11 were written mainly by A.V. Yablokov (with the exception of several sections), Chapters 1, 2, 5 and 8 and some sections in Chapters 3, 4 mainly by S.A. Ostroumov and Chapters 9 and 12 jointly by both co-authors.

Chapters 1, 2, 5 (except for some sections), 8, as well as the Preface, Introduction and some tables were translated into English by S.A. Ostroumov. Chapters 3, 4 and some sections of Chapter 5 were translated by A.S. Vinogradov, and the remaining Chapters by S.V. Ponomarenko. The English translation of the text has been edited and corrected in part by S.A. Ostroumov.

In such a dynamic field as conservation, the situation is changing very rapidly and hence it cannot be excluded that in some cases we were compelled to use information that was either partly out-of-date or not complete. The authors are grateful for all constructive and critical suggestions and comments.

Moscow, January 1991   A.V. Yablokov
                                                                       S.A. Ostroumov

Contents

Part II

Introduction

The 20th century has confronted mankind with many dramatic problems, and the problem of environment deterioration has now become one of the most serious among them. We have not paid sufficient attention to the global long-term consequences of our activities and as a result, our impact on the planet is now so substantial that the vitally important conditions of our existence are being affected and are deteriorating: mankind is faced with a shortage of clean air to breath, a deficiency of unpolluted food to eat, and the absence of the clean water needed to drink.

Increasing global problems — hunger and desertification, depletion of natural resources and deficiency of energy, pollution and man-triggered changes of climate — lead to a better understanding by Man of his dependent position in the Earth's biosphere.

Many of us suffer from the oversimplified illusion that this dependence is obvious and easily understood, and that it is sufficient to respond to environmental problems merely by constructing better waste treatment plants (to have clean air and clean water), by removing mountains of wastes (to have more room for life and recreation), and so on.

However, in reality all this is not sufficient because our dependency upon the biosphere is broader and deeper, and we depend irrevocably upon the diversity of forms of living nature. This is not fully realized by many of us — with the exception of a few biologists. Thus we begin our book with a brief enumeration of the main reasons which make the conservation of living nature so important — no less important than, for instance, the solution of the problems of energy resources.

1. Both the formation and maintenance of the gaseous composition of the air are sustained by living components of the biosphere: primarily by plants and microorganisms, and to a lesser extent by other living beings. Thus, the oxygen so essential for us is being produced by green plants, including forests — boreal, temperate and tropical — and phytoplankton.

2. Soil fertility — one of the essentials for the existence of mankind — is fully determined by the life activities of a great many living creatures in the soil: invertebrates, fungi, bacteria, algae. The number of bacterial cells may reach several tens of billions per 1 g of soil. The total length of fungal mycelium may

reach several km per 1 g of soil. They destroy dead plants and other organisms, convert litter on the soil surface to molecules available for plant nutrition, create a highly heterogenic porous structure of the soil and inactivate toxic substances which may be accumulated in the soil.

3. Purity and quality of water are the result of the activities of living beings. This is an important example of the fundamental principle that all the natural processes of destruction of the ever-increasing amount of pollutants and foreign or natural waste compounds are carried out by biodegradating organisms. Without these, human life is impossible.

4. The diversity of living nature is an indispensable source of all our food (proteins, fats, carbohydrates), top quality materials for clothing (wool, cotton, silk, linen) and foot-wear (leather), cellulose, etc. Half or more of our medicines, including the most widely used, are of natural origin or are made using natural compounds as models. The chemical industry either is not able to replace these products by fully artificially synthesized compounds or can replace them only to a very limited degree. Even in the case of some of these natural products being replaceable, we have no guarantee that their replacement will be without negative effects on Man's health and heredity.

The diversity of forms of living nature gives Man yet another important resource — the resource of genes to improve the characters of agricultural crop cultivars and domestic animal varieties. Thus in the late 1970's in Mexico, a new perennial species of corn, *Zea diploperennis*, was found. It has been used to create new corn cultivars, resulting in billions of dollars profit to US agriculture annually.

5. The potential value of every species is significant. Each species of living nature has its unique characteristics and properties which may be potentially utilized by man now or in the future. At present only less than 5% of plants have been chemically investigated. Annually hundreds of new natural compounds are described which possess pharmacological or biological activities (e.g. natural pesticides and pharmaceuticals including anticarcinogens).

The case of the rainforest plant, the rosy periwinkle (*Catharanthus roseus*), demonstrates how crucially important the potential value of an unexplored species may be. In 1960 a child suffering from leukemia had only one chance in five of remission; now a child has four chances in five, thanks to drugs developed from the rosy periwinkle (Myers 1980c).

Another example of the enormous potential value of a species with all its varieties and strains is the actinomycete, *Streptomyces griseus*: it is difficult to calculate how many useful antibiotics were found among its metabolites, and each year new ones are described.

Unexploited species may offer sources of new sorts of food, building materials, fuel, medicines, natural pesticides and many others (reviews see Myers 1983a; Oldfield 1984).

Methods of biotechnology and genetic engineering greatly enlarge the possibilities of new exploitation of species, in particular of species of fungi, bacteria and algae.

6. Man himself is a sort of ecosystem: each human being is physiologically and physically (as a habitat) connected with an immense number of species of living nature. Our normal process of digestion is impossible without the bacterial species of the intestinal flora: over 10^9–10^{10} bacterial cells (or even substantially more) constantly inhabit the human intestine and produce crucially important metabolites including vitamins, amino acids etc. Some of these metabolites (vitamins) are not produced by our own bodies. In addition, equally numerous living beings inhabit our body surface, contributing to the body's defense against exogenous pathogenic microorganisms.

7. The diversity of forms of living nature is vitally important and essential for the formation, development and maintenance of man's spiritual and mental well-being. Evolutionary man has appeared as a part of living nature and hence ecosystems with numerous and diverse species of animals and plants were for millenia an inevitable part of his surroundings, having a great and favourable effect on both the conscious and subconscious mind. The danger of man being deprived of these beneficial effects is usually underestimated.

8. The position of man in the biosphere and his power over other living beings demand a better recognition of the rights of other species to exist. We must recognize the right to existence of any species — whether it is useful for us or not.

All these reasons, as well as the additional detailed arguments presented by Myers (1983a), Oldfield (1984) and others make the conservation of living nature a matter of great urgency. Unfortunately, we are witness to the process of the decrease of diversity of life on the Earth. One dangerous aspect of this decrease is the extinction of species. Evolution is irreversible, and hence the disappearance of any species means the irreversible loss of a natural resource for mankind. This loss may have even more serious consequences than a loss of oil, gas, coal or metal resources.

In 1982, October 28th, the General Assembly of United Nations accepted the World Charter of Nature. According to the Charter, the genetical basis of life on the Earth must not be endangered, populations of all forms of life must be maintained at a level sufficient for their survival, and biological resources may be exploited only within the framework of their natural potential for reproduction. However, on the path to implementation of these and other essential principles of the Charter there are many obstacles and problems.

The main content of this book is an analysis of these problems (Part I) and of how to solve them (Part II).

Part I

1 The Dangerous Extent of Human Impact on the Biosphere

The degree of anthropogenic transformation of the face of the Earth and of the biosphere has been the subject of numerous works; among them are Ramade 1978, WCS 1980, Ten years after Stockholm 1982, Global 2000 (G 2000), Global future: time to act 1981) and others.

All living organisms, including Man, need some space as habitat; and the surface of the Earth provides habitats for a great many interacting species of microorganisms, plants, animals and fungi, which are integrated into natural complexes (ecosystems); these ecosystems are essential for the long-term survival of Man, for food and oxygen production, for waste product destruction and reutilization, and for the production of many resources vitally important for human well-being and development.

Thus man-made transformation of the surface of the Earth is inseparably connected with the destruction of numerous ecosystems, the disappearance of habitats for species and of the species themselves; the anthropogenic transformation of the land surface is therefore one of the most important aspects of human impact on the biosphere.

Another extremely significant aspect of human impact is the pollution of the biosphere, which is a globally important process with far-reaching consequences for all levels of organization of life systems, as will be discussed in the following Chapters 2–5.

In this chapter we shall briefly consider two aspects: (1) the use and transformation of the land surface, (2) the general level of pollution connected with exploitation of natural resources.

1.1 Use and Transformation of Land Surface

Arable lands with totally transformed ecosystems occupy ca. 11% (estimates differ) of the world's land area, and grasslands used as pastures and thus converted cover an additional ca. 24%. Of the world's land area 31% is considered to be forest land; however, a substantial portion of this area is either man-made forest or forest transformed and damaged due to harvesting or pollution: in some countries more than 30–50% of the forested area is damaged (see Chap. 5).

The total area occupied by roads, industry, and human habitation is up to 2–3% (in some countries even 10% and more) of the land area. Mining works destroy ca. 1% of the terrestrial surface and locally this figure is much higher.

In addition, in many regions man-made lakes and reservoirs have been created, which change the ecosystem of the surrounding lands. Drainage or irrigation works have been carried out on vast areas exceeding the total area of several big countries.

The hydrological regime of huge regions has been changed; this in turn alters the ecosystems. The level of artesian waters in one of the world's largest basins — Ogallala in the USA — as well as under many great cities, has sunk by dozens of meters. The withdrawal of underground water, oil and gas induces land subsidence, which may be over 50 cm per year and reaches, for example, 9 m in the harbour area of Los Angeles-Long Beach, California (Dolan and Goodell 1986). Surface subsidence has affected ca. 22,000 km² in the USA, over 7000 km² in Japan, and many areas in Europe, China, Mexico and other regions. The depletion of underground waters or the planned reduction of their use in order to prevent subsidence in turn stimulates additional consumption of surface waters, which creates changes in hydrological regimes, leading to the further degradation of water body ecosystems.

A substantial portion of important shoreline ecosystems (in densely populated areas even the total shoreline) is being fully transformed. Each year the percent of oil extracted from the shelf near the shoreline is increasing. In the tropics the destruction of mangrove forest ecosystems has reached an extremely dangerous magnitude, with possible negative consequences for offshore fisheries. Two examples of anthropogenic transformation are given in Figs. 1.1 and 1.2.

More detailed information about anthropogenic transmutation and destruction of all the main ecosystem types is given in Chapter 5.

A total of at least 40–50% (in Europe, North America, Japan, and densely populated regions all over the world even more, up to 90% and higher) of the area of the face of the Earth seems to be substantially transformed by Man's activities.

The intense use of biospheric resources (more traditionally called natural resources) is inherently connected with enormous pollution of the biosphere. Some facts about these coupled processes — consumption of resources and pollution — are presented in the next section.

1.2 Exploitation of Natural Resources and Pollution of the Biosphere

In the 1980's, world manufacture of different products was considered to be ca. 2×10^9 t annually. In order to manufacture these products, more than 100×10^9 t of different natural resources have been annually extracted from the biosphere and lithosphere as raw material (French and Spanish equivalents — matières premières and materia prima); that is more than 25 t per year per person. More

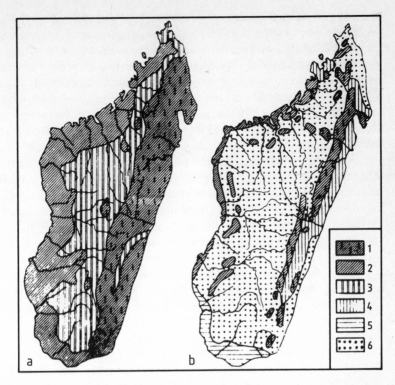

Fig. 1.1a,b. Anthropogenic changes in the ecosystems of Madagascar. **a** original ecosystems; **b** contemporary ecosystems. *1–2* tropical forests; *3* savanna; *4* rice fields; *5* anthropogenic steppes and shrub thickets; *6* pyrogenic pastures. (After Isakov et al. 1980)

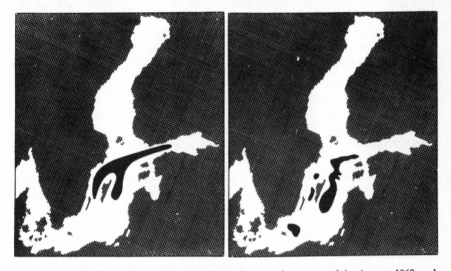

Fig. 1.2. Baltic Sea areas where seabed sediment fermentation occurred in August 1969 and September 1984, releasing hydrogen sulphide. (After Yablokov and Ostroumov 1983)

than 95–98% of raw material is finally excreted and emitted as all sorts of waste material to the environment. Hence the manufacturing of every ton of goods and products is coupled with the introduction to the biosphere of 20–50 tons of different types of waste material. The increased generation of waste material and pollutants is approximately proportional to the growth of industrial production. The following data may give an impression of the rapid growth of pollution of the biosphere: between 1950 and 1983 the industrial production of western developed countries increased 3.8-fold, and that of countries of centrally planned economies of the CMEA 14-fold.

According to estimates based on the summation of different types of pollutants, in industrially developed countries the total quantity of pollutants and different waste products may be more than 1.7 tons per person per year; this figure includes ca. 700 kg of different air pollutants. That is ca. 30 times the total biomass of the whole human populations of such countries (Moore 1985).

In large cities, the production of municipal waste is about 1 t per year per person; the annual average increase of this figure may sometimes reach 3%.

Some data indicate that annually at least 6×10^9 t of different waste materials is dumped into the ocean, i.e., more than 17 t per km^2. Floating refuse may migrate throughout the ocean to distances of thousands of miles.

In large regions of Europe, Asia, and the USA, almost the whole volume of river water flows through industrial, municipal, and irrigation systems.

Estimates exist that annually 18 kg per person of oil and oil products ($65–90 \times 10^6$ t) pollute the biosphere; of this quantity 25% (about 19 million t, mainly lubricants, white spirits, and bitumen) goes into terrestrial ecosystems, and more than 50% (44–68 million t) goes into the atmosphere (calculated from the data given by Miller and Connell 1982). According to satellite data, about 10–15% of the surface of world oceans contains a substantially polluted area.

Among the most important pollutants of the atmosphere are CO_2, sulphur, and nitrogen oxides (Table 1.1), organic compounds etc.

Annual anthropogenic emissions of CO_2 to the atmosphere are higher than 20×10^9 t due to the use of fossil fuels. Moreover, the input of CO_2 to the atmosphere is augmented due to deforestation, phytomass oxidation, and the ploughing up of arable areas, which stimulates oxidation of soil organic matter.

During the 130 years since the beginning of the industrial revolution to 1980, the level of CO_2 in the atmosphere has increased from 265–290 ppm to 338 ppm (i.e. 17% increment) and the rate of further augmentation is at least 0.3% annually. This man-made flow of CO_2 to the atmosphere is believed by some authors to be about 600 kg per capita annually.

The use of fossil fuels is the main reason for emission of S oxides to the atmosphere. The annual increase of these emissions, according to some estimates recently, is 5%. After interaction with H_2O, S and N, oxides give acids which precipitate with the rains (acid rains) back to the land surface. In some states of the USA, the acidity increase in rainwater was 40-fold in 80 years, and ten fold in Italy during the period 1967–1982 (Fig. 1.3).

Annual S deposition over the area of several countries (the Netherlands, Belgium, Austria, Poland, Czechoslovakia, GDR) is about 4 t per km^2 or sub-

Table 1.1. Global pollution by inorganic substances

Pollutant	Anthropogenic production or emissions per year	Natural input to the environment per year	Reference
CO_2	$5-7 \times 10^9$ t C	$2-18 \times 10^9$ t C	1
CO	600–1000 million t	From biota: 420–1120 million t methane oxidation: 400–4000 million t	1
S-containing gases	ca. 65×10^6 t S	Biological processes on the land: 33×10^6 t Biological processes of the sea: 48×10^6 t	2,3
NO_x emissions	19–20 million t N (combustion of fossil fuels)	Biogenic NO_x production: 21–89 million t N	4
Asbestos	4,852,000 t (annual production in 1980)	–	5

Reference: 1 – Smith (1981); 2 – Zehnder and Zinder (1980); 3 – Winner et al. (1985); 4 – Söderlund and Rosswall (1982); 5 – Ney (1986).

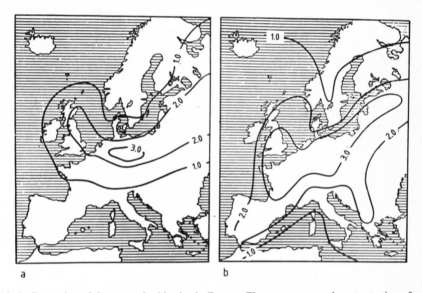

Fig. 1.3a,b. Expansion of the zone of acid rains in Europe. The average annual concentration of sulphur (sulphates, mg l⁻¹). **a** in 1954–1959; **b** in 1972–1979. (Yablokov and Ostroumov 1985)

stantially more. In the FRG depositions of S reach 7 t per km^2 per year (calculated on the basis of data presented by Valenta et al. 1986). In the USA the annual emissions of SO_2 in 1980 were more than 100 kg per person (25,704,000 t) and in the year 2000 it is expected to be at least 26,600,000 t. USA emissions of NO_x in 1980 were more than 90 kg per person (19,293,000 t), in 2000 they will reach 24,100,000 t (projections). The annual increase of NO_x emissions in the world was estimated as 3.5% (Schneider and Grant 1982).

Other globally important air pollutants are ozone and other oxidants which act in combination with SO_2 and NO_x. Ozone monitoring in the FRG, the Netherlands, Great Britain, Denmark, Sweden, and the USA has shown that the former US ambient air standard of 160 $\mu g/m^3$ (80 ppb) was exceeded several times in many measurements in all these countries (Skärby and Sellden 1984). O_3 and other oxidants alone or in combination with SO_2 and NO_x are responsible for up to 90% of the crop loss caused by air pollution in the USA. This crop damage represents an economic loss of US \$1–2 billion.

Man-made flows of many substances to the environment are comparable with the natural ones; sometimes anthropogenic emission is even greater than the natural one (Table 1.2). The fate of many potentially toxic elements and compounds that are being extracted from the lithosphere for industrial purposes is further dissipation in the environment – i.e. the process leading to pollution, as in the case of heavy metals including Pb, Cd, Hg, Cu, Co, Ni, Cr etc.

The global pollution of the biosphere by Pb has rapidly increased during this century and especially after widespread use of Pb alkyl antiknocks in petrols in

Table 1.2. Anthropogenic emissions (E) of several metals to the atmosphere in comparison with natural inputs. (Merian 1984)

Element	$\dfrac{\text{Anthropogenic E}}{\text{natural E}} \cdot 100\%$	Anthropogenic E 10^8 g per year (industrial emissions + fossil fuel)
Fe	39	107,000
Mn	52	3160
Co	63	44
Cr	161	940
V	323	2100
Ni	346	980
Sn	821	430
Cu	1363	2630
Cd	1897	55
Zn	2346	8400
As	2786	780
Se	3390	140
Sb	3878	380
Mo	4474	510
Ag	8333	50
Hg	27,500	110
Pb	34,583	20,300

the 1940's. The Pb concentrations in northern Greenland increased from less than 0.001 $\mu g\,kg^{-1}$ ice in 800 BC, to more than 0.2 $\mu g\,kg^{-1}$ ice in 1969. In the south polar ice of the Antarctic continent Pb was undetectable before 1940 but after that date the Pb concentration was about 0.02 $\mu g\,kg^{-1}$ ice. Some 80,000 t of industrial Pb are present in the upper 50 m of the oceans in the Northern Hemisphere, which is substantially more than the 40,000 t of natural Pb from fluvial sources. Everywhere the level of Pb in the environment and biota is especially high near roads with heavy traffic. For example, in Sydney, Australia, the Pb level in vegetation bordering roads with over 60,000 vehicles per day was 140 $\mu g\,g^{-1}$ biomass, that is seven times higher than near roads with less than 5000 vehicles per day (for review see Bloom and Smythe 1984). In the urbanized areas of the USA, the Pb body burden of the population is 50–1000 times higher than for the inhabitants of the prehistoric period.

Cd was apparently absent in atmospheric precipitations in Western Europe in 1750; in 1930 its annual average deposition was 3.8 g ha^{-1}, and in 1980 5.4 g ha^{-1}. The Cd body burden of the population in industrial countries has increased during the 20th century and at present in some population groups the mean Cd level in the kidney cortex at the age of 50 is only two to five times lower than the critical level for renal damage (Bernard and Lauwerys 1984).

The diversity of dangerous polluting substances is far from being completely estimated. Among such chemicals are organometallic compounds, pesticides [several thousands (Fig. 1.4)], thousands of other organic substances (Table 1.3) with their isomers and congeners. The amount of the latter for PCB's is 209, for PCDF's 135, and for PCDD's 75 (Bandiera et al. 1983).

Pesticides represent persistent and highly toxic pollutants of the biosphere. Per hectare consumption of them is: India 327 g, the USA 1480 g, and Japan about 11 kg (Jalees 1985). Locally, pesticide use may reach 50–200 kg per ha. In areas with heavy use of pesticides, the consumption may attain 5–6 kg per capita annually. The total production of pesticides is over 3 million t per year. The global accumulation of DDT in the environment, according to different estimates, is 1.5–4.5 million t (Lekyavičius 1983), i.e. up to 1 kg per capita. Pesticides and other persistent organic pollutants are being transported through the biosphere by air and waters. The deposition of pesticides from the atmosphere with dust, rain and snow may be as high as several hundred g per month per km^2, even in regions without the use of pesticides.

Over 95–99% of the amount of pesticides introduced into the biosphere serves as a pollutant and poisons non-target organisms, because often only from 0.0000001% to 0.02% of the insecticide applied and 0.1 to 5% of the herbicide applied actually reaches the target pests (Pimentel and Levitan 1986).

The total amount of biocides introduced by Man into the environment is substantially higher than the figures usually reported for pesticides, because additionally almost the same amount of creosote, petroleum, and coal-tar products is used as biocides to preserve wood and other materials. For example, in the USA, almost 500 million kg of pesticides are applied annually (this figure is usually reported) and additionally 500 million kg of creo-

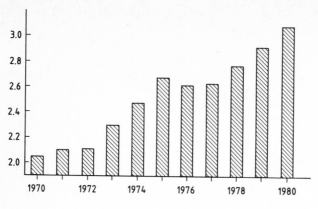

Fig. 1.4. World production of pesticides, herbicides, fungicides and other chemicals to protect agriculture and forestry or used for medicinal purposes. (Yablokov and Ostroumov 1985)

Table 1.3. Global pollution by important organic xenobiotics. (Ostroumov 1986)

Pollutants	Production or emission
Polychlorinated biphenyls (PCB's) Trade names: Aroclor, Phenoclor, Kanechlor, Clophen, Fenclor etc.	Peak annual world production for 1969–1970 70,000 t US production 1930–1975 1.4×10^9 lb Current global burden more than 750,000 t
Polybrominated biphenyls (PBB's)	US production 1970–1976 1.3×10^7 lb
Polychlorinated terphenyls	Peak annual world production 5000 t
Chlorinated naphthalenes	Peak world production of perhaps 5000 t yr^1
Benzene	World consumption as raw material or in gasoline 26–30 million t; input to the environment due to all types of losses ca. 3 million t yr^1
Chlorinated benzene derivatives	World production ca. 900 000 t yr^1
Chlorobenzene	600,000 t yr^1
Dichlorobenzenes	120,000–140,000 t yr^1
Toluene	World production more than 5 million t yr^1; losses during transportation of gasoline 50,000 t yr^1; losses in the chemical industry 50,000–100,000 t yr^1; motor vehicle emission 2 million t yr^1
C_8-aromatics (xylenes, ethyl benzene, styrene)	World production ca. 20 million t
Vinyl chloride	World production 10 million t yr^1, estimated input to the environment due to production losses 200,000 t yr^1
Trichloroethylene (trichloro-ethene)	Input to the environment 600,000 t yr^1
Perchloethylene (tetrachloro-ethene)	Input to the environment 1,100,000 t yr^1
Ethylene dichloride (1,2-dichloroethane)	Input to the environment 1,200,000 t yr^1
Methyl chloroform	Input to the environment 600,000 t yr^1
Detergents (surfactants)	Annual world production of perhaps 10 million t (approximately); US production ca. 2.5 million t yr^1. Input to the environment is practically equal to the production or more, due to technological losses

sote, petroleum and coal-tar products are used as biocides (Pimentel and Levitan 1986).

The adverse action of pesticides and other man-made xenobiotics on organisms, including mutagenic and carcinogenic effects, is discussed below in Chapters 2 and 3.

In addition to industrial pollution, the agricultural pollution of the biosphere due to livestock is immense. One cow produces the same amount of excrement as 16 human beings. The amount of dung-wash and manure that is produced annually in many countries is greater than the municipal water pollution, e.g. in the FRG five times greater, in the USA ten times. In the USA, the waste matter production by livestock is more than 1.7 billion t per year, in the USSR over 1.3 billion t.

Even in the case of excellent treatment of polluted waters the following new problem arises of how to dispose of the sewage sludge produced as a result of the sewage water treatment. This sludge contains toxic halogenated hydrocarbons, heavy metals, PCB's, surfactants and their metabolites. A substantial portion of sewage sludge is used in agriculture as a fertilizer, but this leads to pollution of soils by metals and organic toxic compounds, including chlorinated hydrocarbons, surfactants and their metabolites. The average quantity of sewage sludge produced annually is about 800 kg per inhabitant; this figure does not include industrially produced sludge. In a number of countries the production of sludge is rapidly increasing: 10% per year in Norway, 12% per year in France (Berglund et al. 1984). Industrial dust also threatens the biosphere (Fig. 1.5).

Radioactive pollution of the biosphere is also rapidly increasing. The anthropogenic emission of such radionuclides as ^{3}H, ^{137}Cs, ^{14}C, and ^{239}Pu to the atmosphere is higher — or was higher during a long period of time — than the

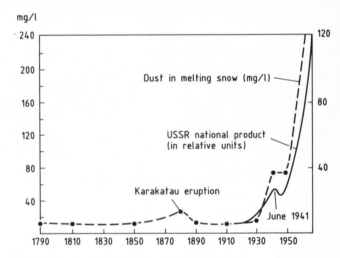

Fig. 1.5. Correlation between the amount of dust settled on the Caucasian glaciers and the volume of industrial production in the USSR. (Yablokov and Ostroumov 1985)

natural emissions. In 1963 the level of ^{14}C in the atmosphere was 100% above the natural level; the reason was nuclear arms testing. Due to nuclear arms production and testing, nuclear power stations, accidents, and coal burning the annual emissions to the atmosphere were, in Ci: ^3H 2.7×10^8, ^{137}Cs 7×10^5, ^{14}C 1.7×10^5, ^{239}Pu 6.1×10^3, calculated on the basis of Jaworowski 1982.

These figures are higher than the natural annual emissions of the radionuclides to the atmosphere, in Ci: ^3H 2×10^6, ^{137}Cs — below detectable level, ^{14}C 3.8×10^4, ^{239}Pu below detectable level. Among dangerous radionuclides is ^{129}I, which is believed — in spite of attempts to protect the environment — to enter global biogeochemical cycles; hence its ionizing radiation emission will influence living beings, including the human population, over many generations (Tikhomirov 1983).

Potentially dangerous, for example in cases of seismic events, accidents or management mistakes, is the storage of huge amounts of radioactive waste over thousands of years.

Thus, by the year 2000, Britain will have accumulated about 4000 t of high-level radioactive waste (RW); about 160,000 t of intermediate-level RW which contains more than 4 GBq (about 100 millicuries) per ton α activity and 12 GBq (300 mCi) per ton β-γ activity; plus about 1 million t of low-level RW which comes not only from nuclear power stations, but also from industry, hospitals and scientific research laboratories (Heath 1985).

By the year 2000, the USA will have about 320,000 m^3 of high-level RW (in 1982 290,000 m^3), 70,000 m^3 of spent fuel (in 1982 7000 m^3), 8 million m^3 of low-level RW (in 1982 2 million m^3), 610 million t of RW produced during uranium ore processing (in 1982 125 million t).

Sea dumping of RW has been extensively practised. During 1967–1976 Belgium, France, the FRG, Italy, the Netherlands, Sweden, Switzerland and the UK dumped in the Atlantic ocean about 46,000 t of RW which contained 4880 Ci of α-emitters and 289,000 Ci of β-γ-emitters. In 1975 the International Atomic Energy Agency (IAEA) considered as "permittable" (!) annual sea dumping up to 10^{10} Ci of α emitters and up to 10^{13} Ci of β-γ-emitters. The total input to the world ocean of ^{238}Pu, ^{239}Pu, and ^{240}Pu appears to be over 230,000 Ci or more than 3 t (Gerlach 1981).

Among the most harmful features of radioactive pollution are: (1) persistence of long-decaying radionuclides in the biosphere; (2) blind non-selective mutagenic or destructive action of ionizing radiation on living beings and cells.

One of the important trends in the development of human impact on the biosphere is the rapid increase of the human population and hence further increases in the consumption of natural resources, transformation of the land surface, and pollution of all sorts, etc.

The total biomass of mankind and livestock was only 5% of the total biomass of all terrestrial animals in 1860, but about 20% in the early 1980's and it will reach about 40% in the year 2020. We are adding 75–80 million human beings a year to our numbers; over a period of 10 years the global population growth rate has been about 17%.

Finally the following points should be stressed.

1. The natural surface of our planet — in Vernadsky's expression, "the face[1] of the Earth" (e.g. Vernadsky 1965) — is being destroyed with an enormous loss of habitats for species and a loss of normal ecosystems which are essential life-supporting systems.

2. Both the magnitude and the growth rate of all sorts of pollution of the biosphere are dramatic menaces to the majority of living organisms, including human beings and the species which are indispensable for human survival and sustaining development.

In order to solve these problems and to manage the situation with living resources in an optimal and economical way, careful analysis is necessary. A detailed presentation of the ubiquitous and dangerous anthropogenic effects on living organisms and ecosystems and of the present trends is given in the following chapters.

---

[1] V.I. Vernadsky did not use the usual Russian word for "face", instead he used the word "lik" (old and poetic Russian word) which is more solemn and broader than a simple "face".

2 Molecular-Genetic Level

Many problems of conservation occur at the molecular-genetic level of life systems. The elementary units at this level are genes represented by segments of DNA (or RNA in some viruses); the elementary events at this level are convariant reduplications of genes. Processes of realization of information stored in DNA involve the functioning of the biochemical machinery of the cell.

A number of serious problems of conservation are caused by the detrimental or disturbing action of anthropogenic factors on molecular systems and processes in the cell.

Another important group of conservation problems are related to different aspects of biochemical transformations and the biodegradation of molecules of polluting substances in organisms and ecosystems.

2.1 Effects on Structural and Functional Systems of the Cell

Among the anthropogenic effects on structures and functions of the cell, the following must be considered separately: (1) effects on genetic systems including mutagenesis, (2) effects on biomembranes, (3) effects on proteins and their metabolism.

2.1.1 Effects on Genetic Systems

The influences of anthropogenic factors on genetic systems may be of several kinds. Thus, direct mutagenic actions of different chemical and physical factors may lead to changes in DNA of gametes and to pathological alterations in subsequent generations. In the case of somatic cells, genotoxic effects may lead to different forms of carcinogenesis in the affected generation.

2.1.1.1 Mutagenesis and Genotoxic Agents
Many of the most common 50,000–70,000 chemical compounds surrounding man in everyday life and polluting the biosphere have genotoxic properties (e.g.,

Klekowski, and Klekowski 1982; Pashin et al. 1983). The total amount of foreign compounds synthesized or introduced to the biosphere by Man is over 4 million; about 1000 new compounds are being introduced to the market annually. One of the results of chemical pollution is that approximately 1950 human beings per 100,000 (about 2%) carry genetically determined pathologies.

Among the main groups of anthropogenic factors which may induce mutations in organisms are the following: ionizing radiation, chemicals used in agriculture, forestry and aquaculture, industry wastes, exhaust from transport vehicles and many others.

For human beings this list may be supplemented by UV-radiation and electromagnetic fields, vibration, medical drugs, food additives, products of smoking, alcoholic beverages (see e.g. Lekyavičus 1983).

Genotoxic environmental pollutants include: PAH; monocyclic aromatic amines; polycyclic aromatic amines and amides; nitro-substituted compounds, e.g. different nitroarenes which are contained in particulate matter in diesel emissions (e.g. Tokiwa et al. 1983); aminoazodyes; aza aromatics; nitrosamines; the majority of pesticides, for instance, Acephate, Captan, Diallat (Garrett et al. 1986), Chlorpyrifos or Dursban (Mostafa et al. 1983), carbamate pesticides including carbaryl (α-sevin, β-sevin), zectran (Mexacarbate), benomyl (Benlate) (Woo 1983) and many others; many metal compounds containing Cr, Sb, As, Cd, Co, Cu, Pb, Mn, Hg, Ni, Tl, and numerous other substances (Wildemauwe et al. 1983; Williams 1985; Kier et al. 1986; Kolman et al. 1986, etc.).

A recent review of tests of 1078 chemicals in *Salmonella* has shown that 603 (56%) were definitely mutagenic; for about 327 (30.3%) no conclusions could be drawn, and only 148 (13.7%) were non-mutagenic. Moreover, several groups of chemicals are known to contain a high proportion of false non-mutagens in the *Salmonella* assay; among such important genotoxic chemicals are azo compounds, carbamyls and thiocarbamyls, halogenated compounds, some inorganic compounds etc. (Kier et al. 1986).

Several international and national lists of environmental mutagens have been compiled, which include over 10,000 different compounds. An additional indication of the enormous genotoxic potential of pesticides is the following: it has been shown that the frequency of chromosome aberrations in lymphocytes of some categories of the rural population is significantly higher in the areas where pesticides are intensively used.

In addition to the chemical mutagens, several physical factors may also induce mutations, among which ionizing radiation and electromagnetic fields are very important. Numerous experiments with microwaves have shown their ability to induce aberrations of chromosomes in lymphocytes of mammals, including man. Dominant lethal mutations have been induced in mice by an electromagnetic field with a frequency of 2.45 GHz and a density of energy flow of 50–100 mW cm^{-2}. Radiofrequency radiation (27 MHz) exposure caused an increase in chromosome breaks in mononuclear leukocytes (Roberts et al. 1986).

The simultaneous action of several mutagens may substantially enhance their influence. However, the combined action of mutagens or potential mutagens has not been investigated extensively. Such studies are of great ecological interest.

It has been shown that specimens of water and silt from many water bodies may induce mutagenic effects. Thus, the mutagenic action of water has been demonstrated for: the Mississippi River and Lake Ontario (mutagenicity for *Salmonella typhimurium*), the Rhine (mutagenicity for the fish *Umbra pygmaea*), the Danube (mutagenicity for *S. typhimurium* and *Drosophila melanogaster*). The silt of many European water bodies was shown to be mutagenic for plants (Kurinny 1985).

It seems possible that water mutagenicity is the result of combined (sometimes synergistic) action of several compounds. This stresses again the gaps in studies on the mutagenic action of mixtures of several pollutants.

One of the obstacles in the study of mutagenic and other biological effects of environmental pollutants is the high cost of a full investigation of every polluting substance using a wide range of test organisms.

As conclusion to this section devoted to environmental mutagens, it is worth generalizing that at present both global and local pollution of the biosphere and its components is creating a dangerously high level of mutagens in the environment and habitats for many organisms.

2.1.1.2 Extrachromosomal Genetic Determinants

Another kind of anthropogenous effect on genetic systems which is extremely important for conservation problems is the pollution of the biosphere by extrachromosomal genetic determinants including: (1) plasmids, (2) artificially synthesized recombinant DNA, (3) viruses.

Plasmids. Many regions of the biosphere are being polluted by the plasmids that contain genes (factors) of resistance for antibiotics.

Thus, in Japan the frequency of isolation of resistant bacterial strains of *Shigella* has increased many times during recent decades. The frequency of isolation of the strains resistant to at least one of several antibiotics was the following: in 1952 — 0.04% of all investigated strains; in 1972 — 90.5%. Many strains of *Salmonella* have shown resistance to streptomycin, tetracycline and other antibiotics in the study by Tanaka et al. (1976).

In 1983, plasmid pollution of 848 bacterial strains of the family Enterobacteriaceae that were isolated from Lake Champlain and Mud Hollow Brook (Vermont, USA) was investigated by Mietz and Sjogren (1983). Of the isolates tested 87% were resistant to six or more antibiotics. The percentage of total isolates resistant to individual antibiotics was the following: cloxacillin 98.9%, penicillin 98.8%, clindamycin 98.5%, novobiocin 96.7%, ampicillin 76.4%, carbenicillin 75.9%, cephalothin 49.5%, erythromycin 48.0%, chlorotetracycline 29.0%, tetracycline 14.6%, streptomycin 9.3% etc. The genera isolated most often were *Klebsiella*, *Escherichia* and *Enterobacter*. Mietz and Sjogren have confirmed that the property of antibiotic resistance was plasmid-mediated and could be transferred to recipient cells of *E.coli* via conjugation or transformation. The resistant bacteria were capable of persisting in the environment.

The enormously broad distribution of plasmids in recent years has been stimulated by: (1) use of antibiotics as medicinal drugs, (2) addition of antibiotics to feed. Thus, about 40–45% of the antibiotics produced in the USA is being used as an addition to feed. In 1983 5.4 million lb of antibiotics were administered to animals in the USA. Similar additions are made in other countries also, and the pollution of food products by antibiotics is a common phenomenon. Thus, in one of the studies in Italy it was found that among 50 samples of milk 16 contained chloramphenicol, 15 tetracyclin, 13 streptomycin. Moreover, 13 samples were polluted by two or three antibiotics simultaneously (Aureli et al. 1975).

Among the numerous demonstrations of antibiotic resistance in bacteria are the works on *Salmonella, Shigella, Serratia, Staphylococcus, Pseudomonas, Aeromonas*, coliform bacteria and many other taxons.

A wide range of antibiotics is also being used in aquaculture for the control of bacterial and fungal diseases. These compounds are being used as food additives, baths and as injections. They may induce antibiotic pollution and resistance in aquatic microflora, as has been shown for four sites producing rainbow trout (*Salmo gairdneri*) in Great Britain, where resistance of fish farm bacteria (*Acinetobacter calcoaceticus, Aeromonas hydrophila* etc) to penicillin G, chloramphenicol, erythromycin, streptomycin and 14 other antibiotics has been detected (Austin 1985).

An additional source of plasmid pollution in some regions could be fish exported from Southeast Asia to North America and possibly to Europe. It has been shown that the main component of microflora of aquarium fish from Southeast Asia was *Aeromonas hydrophila*; the cells of *A. hydrophila* isolated from aquarium fish and waters contained R factors that determined resistance to tetracyclin, streptomycin, ampicillin, chloromycetin (Shotts et al. 1976). The possible reason for the occurrence of R factors in the bacteria is the prophylactic addition of antibiotics to aquarium water.

The results of broad plasmid pollution in the biosphere are far from being fully understood. At the very least, it is not desirable, because this sort of pollution makes the use of antibiotics in medical treatment more difficult and sometimes impossible.

Another negative result of plasmid pollution is the creation of potential new ways for the transfer of genetic determinants from one bacterial cell to another or the transfer of genetic determinants through the chains: chromosome-phage-plasmid or plasmid-chromosome-plasmid-plasmid etc.

The possibility of the transport of genetic elements and the ability of plasmids to integrate with chromosomes may lead to the growth of instability of bacterial genetic material, induced by anthropogenic dissemination of plasmids in the biosphere. This phenomenon causes much trouble.

Anthropogenic Recombinant DNA. The boom in genetic engeneering means that in many laboratories hybrid DNA molecules exist, which include in their structures fragments of DNA of different species. These fragments are linked together in vitro. On this basis the possibility of anthropogenic synthesis of new chimerical organisms exists, and all biological properties of such organisms cannot be fully predicted.

In a number of experiments, the recombinant DNA molecules were introduced to plasmids or phages of *Escherichia coli* K 12. In some experiments, 10^7–10^9 bacterial cells of *E. coli* K 12 have been introduced to man, and 1 day after this treatment 1 g or faeces of the subject contained 10^3–10^8 living cells of *E. coli* of this strain. The conclusion may be made that the significant amount of cells with plasmids and/or recombinant DNA fragments may come out of the laboratories and enter the environment.

The addition of antibiotics to feed and other numerous cases of the presence of antibiotics in the animal and human environment may give selective advantage to those bacteria that have plasmids with recombinant DNA. This in turn could lead to an increase in the percentage of such cells in the bacterial flora of organisms.

During their evolution, animals and man have never faced such an unusual challenge as the presence in their digestive system of new compounds which may be produced by bacteria carrying recombinant DNA.

Viruses. The pollution of the biosphere by viruses is a new and growing type of pollution by extrachromosomal genetic determinants.

The concentration of population, in urbanized areas creates especially favourable conditions for the circulation of many types of viruses. Some of them may incorporate in their circulation not only man, but different other organisms also, including both wild and domesticated animals.

Thus, the isolation of influenza A viruses in the USSR has been reported for:

1. many wild animals including the birds *Larus ridibundus, L. genei, L. ichthyaetus, Erithacus rubecula, Anas crecca, A. querquedula, Tetrastes bonasia, A. clypeata, A. acuta, A. falcata,* several species of Corvidae, *Streptopelia turtur, Motacilla alba, Sterna hirunda, S. albifrons, S. paradisaea, Esox lucius, Bucephala clangula, Hirundo rustica, Nycticorax nycticorax* etc., mammals *Capreolus capreolus, Sciurus vulgaris,* several species of Balaenopteridae, *Nyctalus noctula, Rangiter tarandus* etc., fishes *Oncorhynchus* sp.;
2. domestic animals — turkey, pigeon, hen, duck, pig, dog calf (Lvov and Zhdanov 1983 and others). A 5- to 25-year persistance of genes coding haemagglutinins of all known epidemic human viruses (H1, H2 and H3) in the gene pools of natural populations of viruses have been established. Lvov and Zhdanov (1983) supported the possibility of exchanges between the gene pools of natural populations of influenza A viruses and epidemic influenza viruses.

Among anthropogenic factors of importance for virus pollution are the following: (1) concentration of domesticated animals in livestock farms; (2) immigration of people and domesticated animals into new sites where natural ecosystems harbour local aborigenous virus circulation (in Russian scientific literature such ecosystems are called "natural hotbeds" of aborigenous virus infections); (3) emigration of population from urbanized areas to the countryside; the migrating people may carry with them those virus strains that circulated in the cities.

These broad migrations accelerate the flow of genetic determinants of viruses from one point of the biosphere to others. The results are the creation of new forms and strains of viruses, epidemics and pandemics.

In addition to terrestrial environments, the aquatic ecosystems are also being heavily polluted by viruses. More than 110 human viruses are found in sewage and are collectively known as "the enteric viruses" (Goyal 1984); among them are 37 types of adenoviruses, 31 types of echoviruses, 23 types of coxsackieviruses A etc. They cause respiratory desease, eye infections, meningitis, diarrhoea, rash, fever and many others. The average enteric virus concentration in domestic sewage is about 7000 viruses per liter, but as many as 460,000 viruses per liter have been found in some parts of the world. Per g of faeces 10^9 virus particles may be excreted. About 8 billion gallons of municipal sewage are being discharged each day into the coastal waters of the USA, only half of it receiving secondary treatment. Even disinfection by chlorination may not inactivate all viruses (Goyal 1984).

The after-effects of interactions between viruses and host genomes are not understood completely. Possible consequences include: (1) mutagenesis (which may be induced during the extraction or insertion of DNA fragments), (2) enhancement of action of chemical genotoxic pollutants, (3) vice versa, potentiation of viral action by chemical pollutants (carcinogens), which has been shown in several in vivo and in vitro systems.

The examples of virus-induced mutagenesis are the lethal mutations in chromosome 2 of *Drosophila melanogaster* induced by several RNA viruses (Alexandrov and Golubovsky 1983) and mutations in clones of murine fibroblasts treated by virus SV-40 by other authors. Several other examples were given by Gershenzon et al. (1975).

Effects of enhancement of action of genotoxic pollutants when they act together with viruses may be so significant that on this basis a special bioassay system for screening suspected genotoxic environmental pollutants has been developed. The following classes of chemical compounds were able to enhance the adenovirus transformation of Syrian hamster embryo cells: alcohols, phenols, aliphatic amines, alkyl sulfates, aromatic amines, aryl halides, carbohydrates, hydroxylamines, mycotoxins, polycyclic hydrocarbons, hydrazines, metal salts, steroids and chlorinated hydrocarbons.

Hatch et al. (1983) have studied several chlorinated methanes and ethanes, including 1,2-dichloroethane (the annual production of 1,2-dichloroethane is about 12 billion lbs in the USA alone), and have shown that the volatile liquids 1,2-dichloroethane, 1,1,1-trichloroethane, dichloromethane, and chloroform substantially enhanced transformation of Syrian hamster embryo cells by SA 7 adenovirus. The gases chloromethane and vinyl chloride were also active in this biotest system.

Other authors have demonstrated that 9,10-dimethyl-1,2-benzanthracene (DMBA) and benzo(a)pyrene accelerated the incidence of erythroleukaemia induced by the Friend leukaemia virus (FLV) in SJL/J and B10SJF1 mice. Those mice that received both virus and chemical died sooner than the animals that received either virus or chemical alone (Raikow et al. 1983).

Full examination of virus-mediated anthropogenic effects in the biosphere must include also potential hazards connected with experiments involving retroviruses and xenotropic viruses.

In conclusion, different anthropogenic factors (pollution by mutagens, plasmids, viruses, and possibly by recombinant DNA) may enhance the mutagenic process and disturb genetic processes in the biosphere. A close connection may occur between the mutagenic ability of anthropogenic factors and their cancerogenicity. The latter type of effect is cosidered in the next chapter, on the ontogenetic level of life systems.

2.1.2 Effects on Biomembranes

The molecular mechanism of action of many pollutants on the living cells is connected with detrimental effects on the structure and function of biological membranes.

Numerous anthropogenical factors disturb biomembranes: many heavy metals, gases, organic compounds (including pesticides and detergents), electric fields and ultrasound.

The production of detergents in the world has increased from 30,000 t in 1949 to several million t. Thus, the USA produces more than 2 million of detergents annually. The annual consumption (and discharge to the environment) of detergents is substantially more than 4 kg per capita (Hutzinger 1982).

The following important types of effect on the biomembranes should be mentioned.

2.1.2.1 Effects on Permeabilities and Ion Channals of Biomembranes

This type of effect has been shown, for example, for many pesticides, including organochlorines (DDT and many others), dithiocarbamates, nitrophenols (for instance, dinitrophenol, a well-known uncoupler in membrane bioenergetics), derivatives of urea (e.g. Ashton and Crafts 1981; Narahashi 1982; Popova Profirov et al. 1982; Ziegler 1982; Corbett et al. 1984). A similar type of action has been demonstrated during study of the action of several polluting gases (SO_2, O_3, nitrogen oxides), metals and metallorganic compounds (e.g. Elkiey and Ormrod 1979; Kawa 1979; Mukherji and Mukerji 1981; Byczkowsky, Sorenson 1984).

It seems possible that these pollutants (especially pesticides) which uncouple phosphorylation in coupling membranes act as agents that increase the membrane permeability. There are many examples of the uncoupling action of different pesticides (for instance Wright et al. 1980, Ashton and Crafts 1981 and others).

It is interesting that recently it has been shown that such antropogenous factors as infrasound and low-frequency noise may change the permeability of erythrocyte membranes.

It is worth stressing that the changes of membrane permeability to ions may induce shifts in the magnitude of transmembrane electrochemical potential and of surface charge densities. Both these parameters of the biomembrane are important elements of the cell regulatory system (Ostroumov and Vorobiev 1976; 1978; Ostroumov et al., 1979) and hence the changes of these parameters may have unpredictable and detrimental implications for the cell. Several new findings are in accordance with this conclusion.

Direct evidence of effects on the magnitude of membrane potential of several pollutants and their analogues has been obtained in some experiments, for instance during the investigation of the effect nitrosamines on animal cells, of Ni ions on the roots of *Zea mays* and in other works with organic xenobiotics (Imaida et al. 1983; and others).

2.1.2.2 Effects on the Chemical Composition of the Biomembrane

This type of effect is frequently demonstrated when the broad spectrum of compounds is being biotested. Among these effects two important groups may be discriminated: (1) the inhibition or other disturbance of biosynthesis of the molecules of biomembranes; (2) the peroxidation of membrane lipids. We shall consider these effects separately.

1. *Effects on Biosynthesis of the Membrane Molecules.* Compounds of practically all types of chemical components have been found to be able to inhibit the normal biosynthesis of biomembrane molecules. Especially numerous are the indications of such effects in different pesticides which disturb the biosynthesis of fatty acids, carotenoids, chlorophylls, proteins.

Among the pesticides (especially herbicides) which poison the lipid biosynthesis are haloalcanic acids (e.g. potassium trichloracetate and Dalapon), nitriles (including Ioxymil, Dichlobenil), anilides (e.g. Alachlor, Metolachlor) nitrophenols (among them Dinoseb), thiocarbamates (EPTC, Diallat, Sulfallat, Pebulat, Vernolat), heterocyclic compounds (for example Ethofumesat and many others) etc. (Mudd et al. 1971; Muslih and Linscott 1977; Grumbach and Bach 1979; Rivera and Penner 1979; Ashton and Crafts 1981; and others).

Detergents have been shown to have the ability to change the fatty acid composition of the plant cells. Several detergents studied (Triton X-100, sodium dodecyl sulphate, sodium desoxycholate and cetyl trimethylammonium bromide) increased the normal ratio between mono- and digalactosyl diglycerides in red algae *Porphyridium purpureum* (Nyberg and Koskimies-Soininen 1984).

Some gases polluting the atmosphere may affect biosynthesis of molecules of biomembranes. Among them ozone, which may inhibit the biosynthesis of glycolipids in plant chloroplasts and induce lipid changes in the plant cell (e.g. Mudd et al. 1971). Heavy metals may also cause alterations of lipid biosynthesis.

2. *Effects on Peroxidation of Membrane Lipids.* The peroxidation of lipids is an extremely important and potentially dangerous process which is stimulated by many pollutants, including different organic substances and gaseous oxidants (ozone and others).

This type of effect has been discovered under the influence of such substances as the derivatives of urea, triazines, dipyridins, galogenated phenoxyacids, CCl_4 and many others.

It should be emphasized that the origin of new peroxy groups in the membrane may increase the permeability of these parts of the biomembrane. The result will be as has been discussed above.

Moreover, the rapid generation of toxic products of peroxidation of lipids in the cell may: (1) exaust the resources of natural antioxidants, (2) decrease the activity of enzymes which contain thiol groups, (3) decrease the K^+ and Na^+ transport across the membranes which is being fulfilled by Na, K-ATPases.

The peroxidation of lipids is associated with the generation of free radicals which may exhibit detrimental effects on the molecular structures of the cell.

The toxicity of one dangerous pollutant — 2,3,7,8-tetrachlorodibenzo-p-dioxine (TCDD) — is connected with the increase of in the peroxidation of lipids in the hepatic cells of rats studied in the experiment (Stohs et al. 1983). This finding adds important information to the previous data about the action of this compound on ATP generation in mitochondria, on the biosynthesis of proteins and lipids, on the metabolism of glucose, alanine, oleate etc.

2.1.2.3 Photodestruction of Chlorophylls in Photosynthetic Membranes

The photodestruction of chlorophylls may take place due to the interruption of the processes of energy transfer from the chlorophyll molecules which have absorbed the quanta of light energy, to the acceptor molecules.

The same result may occur due to another reason, e.g. the abolishment of the protecting action of carotenoids which may prevent the photooxidation of chlorophyll molecules. It has been mentioned above that some pollutants, including pesticides (among them derivatives of pyridizinone), block the biosynthesis of carotenoids and through this mechanism may induce photooxidation of chlorophylls in tissues of plants.

2.1.2.4 Detrimental Effects on the Ultrastructure of Membraneous Organelles

Numerous disturbances of this type have been revealed in the course of ultramicroscope studies of chloroplasts, mitochondria and other organelles.

In chloroplasts changes in ultrastructure have been discovered following the action of pesticides (derivatives of urea, thiadiozinone, pyridazinone and other compounds), some metals and gases (HCl, SO_2 and others). Observations of such effects are extremely numerous (e.g. Guderian 1977; Ashton and Crafts 1981; Byczkowski and Sorenson 1984).

2.1.2.5 Molecular Interactions of Pollutants with Membranes Components

It seems appropriate to mention also an impressive body of additional data on the direct interaction of different xenobiotics, including pollutants, with molecular constituents of biomembranes.

Thus, it has been shown that organochlorine pesticide chemicals may block the energy supply to the coupling membrane of mitochondria. Using fluorescent probe, 1-anilino-8-naphthalene sulfonate, it has been demonstrated that such pesticides as DDT, Methoxychlor and Kelthane reduce the supply of energy from succinate and ATP to the inner membrane (Hijazi and Chefurka 1982). This finding supports numerous previous indications that DDT interacts with molecular mechanisms of energy transduction in coupling membranes of mitochondria.

In accordance with these data are the results obtained when model membranes of lipid vesicles have been studied. Using the thermal dependence of fluorescence polarization of 1,6-diphenylhexatriene, the differential effects of DDT and pentachlorophenol on lipid bilayers of dipalmitoylphosphatidyl-choline liposomes have been shown by Buff et al. (1982).

Other works have also demonstrated the direct interaction of organic and inorganic pollutants with the molecules of biomembranes or model membranes.

The experimental evidence of the interaction of pollutants with membrane proteins is considered below in the section devoted to proteins and enzymes.

Metal-containing environmental pollutants may interact with molecular components of biomembranes. One example of evidence in support of this is the study of the effects of methylmercury on agglutination of rat sarcoma cells (Nakada et al. 1981). In this work the agglutination of Yoshida ascites sarcoma cells which were harvested from donor rats was significantly depressed in the presence of 10^{-6}–10^{-5}M methylmercury.

Metal cations (La^{3+}, Pb^{2+}, Cu^{2+}, Cd^{2+}, Zn^{2+} and others) may interact with biomembranes. Their bonding on the outer surface of the biomembrane may decrease the net negative charge and disturb normal functioning of the membranes (for review see Ostroumov and Vorobiev 1978, Ostroumov et al. 1979; Byczkowski and Sorenson 1984).

The accessibility of heavy metal compounds to the hydrophobic zone of the membrane is an important question. The interaction of mercury compounds with the hydrophobic portion of the membrane has been studied, for example, by measuring the fluorescence quenching of a hydrophobic probe which was embedded into the phospholipid bilayer. Using fluorescent quenching of the hydrophobic probe pyrene, the accessibility of mercury compounds ($HgCl_2$ and CH_3HgCl) to the hydrophobic core of artificial liposome membranes made of phosphatidylcholine and phosphatidylserine has been shown by Boudou et al. (1982).

Some new facts have been found during the investigation of action on the cell bioenergetics by acceptation of electrons from electron transport chains. Thus the herbicide Paraquat (methyl viologen) is able to accept the electrons from redox carriers of chloroplast or microsome membranes. As a result the molecule of Paraquat is reduced to its corresponding cation radical. Later, in the course of further reactions of cyclic autoxidation, the superoxide anion radical is produced. The significant degree of toxic effects of Paraquat may be dependent upon the increase in production of this superoxide anion radical (e.g. Farrington 1976; Klimek et al. 1983).

Generalizing the broad amount of data it should be stressed that in many situations it is impossible to indicate exactly the primary or main mechanism of the action of pollutants on the cells. This is often true in the case of the pollutants which disturb the transformation of energy in biomembranes. It is often difficult to say which effect is primary: the change of membrane permeability or the interruption of accumulation of energy in the macroergic bonds, first of all in the ATP.

Finally we emphasize the possibility of complex interconnection between three types of pollutant effects — disturbance of biomembranes, influences on the genetic systems and on the protein/enzymatic systems. For instance, such pesticides as amiprofosmethil, trifluralin and oryzalin induce the release of calcium from mitochondria to the cytoplasm. The increase of calcium concentration in the cytoplasm may inhibit the formation of protein microtubuli, and hence the formation of the mitotic spindle. The final result will be the blocking of chromosome movement during mitosis. It is clear that the possibility of such interconnections may make understanding the effects induced by pollutants considerably more complicated.

2.1.3 Effect of Pollutants on Proteins/Enzymes

An enormous number of substances inhibit the functioning of proteins which act as redox carriers in biomembranes, and the result is damage to cell energetics. For instance, many pesticides may cause effects of such type, among then the derivatives of urea, quinones, S-triazine and phenolic herbicides, Dinoseb, Rotenone DDT, Hydromethylnon etc. (e.g. Akbar and Rogers 1985; Hollingshaus 1987; Oettmeier et al. 1987). Other organic substances and heavy metals may also inhibit redox chains.

It is interesting that some weed plants are resistant to the herbicide atrazine, which normally also inhibits the redox chain in chloroplasts. The resistance is dependent upon the modification of the chloroplast membrane protein which is bound to the atrazine molecule in non-resistant plants. The inability to bind atrazine results in the absence of interaction between the atrazine molecule and the protein, and this herbicide does not inhibit electron transfer through the redox chain in resistant weeds.

The interruption of electron transfer by pollutants may have multiple consequences. Thus the chlorophyll molecule which cannot donate absorbed energy to the redox chain may proceed to photodestruction. This leads to the loss of chlorophyll by the plant, and finally chlorosis may occur.

Another extremely important aspect of cell energetics is oxidative phosphorylation, i.e. ATP synthesis. This process may be harmfully influenced by many organic xenobiotics, e.g. DNOC (dinitroorthocresol), Dinoseb and some others.

Of outstanding importance are the effects of many pollutants on the activity of microsomal enzymes which are responsible for: (1) metabolism of steroid hormones, fatty acids, alkylpurines, tyroxine and bioactive amines; (2) biotransformation of xenobiotics.

Among the substances able to inhibit or change the normal functioning of hepatic microsomal enzymes are different pesticides, CCl_4, CS_2, CO and others.

An important type of change in the activity of these enzymes is the induction of microsomal enzymes under the influence of organochlorines, polyaromatic hydrocarbons, several metals (Ni, Cr, Cd, Pb) and other toxicants. Thus it has been shown that the pesticides Chlordan, DDT, Dildrin and also PCB's may lead to acceleration of hydroxylation of steroid hormones (androgens, oestrogens and glucocorticoids) in several species of vertebrates. The result may be the accelerated disactivation of these hormones. The decrease in the level of steroid hormones in the organism (caused by the induction of microsomal hydrolases under the action of chlorinated or brominated hydrocarbons or other pollutants) may in turn lead to inhibition of reproduction in the animals.

The pollutants may change the activity of enzymes bound to other membrane structures.

One extremely important type of compounds which disturb protein functioning is the group of organophosphates; these substances may inhibit acetylcholinesterase.

An important class of effects of pollutants on proteins is the interaction of some xenobiotics with microtubules (Zimmermann et al. 1985). Thus, in the University of Rochester (NY, USA), it has been shown that methylmercury (MeHg) caused disruption of cellular microtubules in monolayer cultures of non-transformed human fibroblasts (Sager et al. 1983). In vitro polymerization of pig brain microtubule protein was inhibited by the addition of MeHg.

This may at least in part explain why MeHg is known to inhibit mitosis in cell culture systems, and why MeHg is toxic for the developing brain. MeHg is known to be both a potent neurotoxic agent and a teratogen in mass poisoning in Japan and Iraq.

Some pollutants may interact with extremely important protein calmodulin which plays a key role in many calcium-mediated cellular processes. At the Friedrich Miescher Institute (Basel, Switzerland) it has been shown that DDT at concentrations found in living organisms competitively inhibits calmodulin-mediated activation of $3',5'$-cAMP phosphodiesterase from calf brain (Hagmann 1982). Hagmann has concluded that inhibition of calmodulin-mediated mechanisms might therefore be an important side effect of DDT.

It is known that the formation of avian egg shell requires the supply of carbonate ions which are involved in the production of calcium carbonate. A key enzyme in these processes is carbonic anhydrase. It has been shown that DDT (through its metabolite DDE) and PCB's may inhibit carbonic anhydrase activity and hence induce thinning of the egg shell. Thus, DDE-induced egg shell thinning in a highly sensitive species, the American kestrel (*Falco sparverius*) was associated with significant inhibition of carbonic anhydrase (and also Ca-ATPase) in the shell gland (Bird et al. 1983).

Among a great many other enzymes that are inhibited by pollutants are those participating in synaptic transfer of signals, as well as other vitally important enzymes including:

— LDH, which was inhibited by s-triazine pesticides and PCH's (Barthova et al. 1985),

— α-chymotrypsin, which was inhibited by chlorinated hydrocarbons, PCB's and alkanes (Law et al. 1985).

It is necessary to emphasize that pollutants may inhibit those enzymes which participate in the biosynthesis of practically all the components of the living cell.

In the previous section we have given examples of the inhibition of biosynthesis of lipids that were components of biomembranes. Another important set of facts is the sensitivity of aminoacyl-tRNA synthetases to the xenobiotics attacking thiol groups of proteins.

Pollutants may not only inhibit enzymes or their biosynthesis but in many cases stimulate them. One especially important class of effects of this kind is the stimulation of microsomal enzymes involved in xenobiotic oxidation, including the uniquely important and polyfunctional P-450 system. Stimulation of other enzymes has also been described.

Thus, an important fungicide and a byproduct in many industrial processes is hexachlorobenzene (HCB) which increases the activity of gamma-glutamyl transferase (EC 2.3.2.2) in the liver and in the serum of rabbits and rats. At the Institute of Gastroenterology and Nutrition (Medical Academy, Sofia) it has been shown that, following a 60-day exposure, the enzymatic activity in liver exceeded control values by more than 60 times. (Adjarov et al. 1982).

Organochlorine pollutants may cause significant increase in the activity of other transferases and of enzymes of other classes. Thus, in Nagoya University (Japan) it has been shown that PCB's, DDT and other xenobiotics increased the activity of hepatic UDP-glucuronyl transferase in rats 2–13-fold. In the same work it was demonstrated that these xenobiotics also stimulated the activity of hepatic UDP-glucose dehydrogenase in rats 1.3–2.8-fold (Horio et al. 1983).

It is possible that such data reflect the functioning of defensive mechanisms of organisms and cells; however, sometimes these "defensive" biotransformations may produce toxic compounds (see Sect. 2.2 of this chapter).

In the literature devoted to the mechanisms of action of halogenated pollutants, the possibility is discussed that, like dibenzo-p-dioxins (PCDD's, including 2,3,7,8-TCDD's), the PCB's, polychlorinated dibenzofurans (PCDF's), and polybrominated biphenyls (PBB's) affect the cells and organisms through a cytosolic receptor protein that preferentially binds the toxins which are approximate isostereomers of 2,3,7,8-TCDD (Safe 1985; 1986; Bandiera et al. 1983). Correlations have been shown between the toxic potency of PCDD's and their potency to induce cytochrome P-448-dependent aryl hydrocarbon hydroxylase (AHH), between the potencies of individual PCDD congeners to induce AHH activity and their affinities for receptor binding.

Recent data indicate the importance of cell enzymes involved in the formation of the level of superoxide anion generation and of lipid peroxidation, i.e. superoxide dismutase and catalase. Both these enzymes may be inhibited by the action of xenobiotics on organisms, as was shown when Lindane was administered to rats by Junqueira et al. (1986).

2.1.4 Other Types of Effects of Pollutants.
Interconnection of Effects on Different Molecular Structures

Acting through different mechanisms, pollutants may change the normal concentrations of substances that are vitally essential for metabolism, bioenergetics, and their regulation: ATP, ADP, inorganic phosphate, phosphocreatin, calcium, magnesium, hormones etc.

Thus, this type of effect has been demonstrated by Kopp et al. (1983) in rats following chronic low-level cadmium (or cadmium + lead) exposure approximating environmental heavy metal levels. In this experiment rats received cadmium and lead in drinking water (1 μg/ml) from weaning to 18 months.

Changes in normal levels of hormones and other important molecules in the blood of animals may be induced not only by xenobiotics but also by physical factors. Thus, radio frequency radiation may increase corticosterone concentrations and decrease concentrations of thyrotropin and growth hormone (Roberts et al. 1986). Very low frequency electromagnetic field may reduce calcium uptake by stimulated lymphocytes (Conti et al. 1985) and this may indicate a serious disturbance of normal molecular processes in these important cells.

In many cases it is extremely difficult to identify the primary molecular target in the molecular mechanism of action of a given pollutant.

The inhibition of biosynthesis of RNA and/or DNA will automatically lead to disturbance of protein biosynthesis. Reciprocally, the inhibition of the activity of enzymes participating in the biosynthesis of nucleic acids will cause disturbances in the metabolism of the latter.

It appears therefore quite natural that many pollutants disturb the metabolism of both proteins and nucleic acids. Among such pollutants are, for example, almost all the main classes of herbicides (e.g., Corbett et al. 1984).

Any agent that interacts with the biomembranes and changes their permeabilities, membrane potential, and/or surface charge densities may induce a breakage of the bioenergetics of the cell or of the cell regulatory machinery (Ostroumov and Vorobiev 1976, 1978; Ostroumov et al. 1979) with further ramifications for the metabolism of proteins and nucleic acids.

The interconnection of effects induced by pollutants, the possibility of the existence of cascade- or trigger-type mechanisms of action of xenobiotics and anthropogenic factors make it enormously difficult to predict the full range of biochemical effects that may be produced by pollutants (see also Chap. 8).

In conclusion we can generalize that many pollutants are often nonspecific in their action at the molecular level and are frequently able to cause a broad spectrum of effects connected with rupture, perturbation, and disorder in genetic, membrane, and protein/enzyme systems.

2.2 Biotransformation and Biodegradation of Pollutants

The popular synonym for foreign compound (xenobiotic) metabolism is detoxication; but there are cases where the metabolism achieves no detoxication or even involves a reverse effect. One and the same type of metabolic pathway may lead to either a decrease or an increase in the toxicity of xenobiotics, so that it is best to refer to all these reactions as biotransformation.

Biotransformation of pollutants and of xenobiotics usually involves the following types of enzymatic reactions: (1) oxidation, (2) reduction, (3) degradation, (4) conjugation.

A prominent role in oxidoreductions of xenobiotics is played by cytochromes P-450, a family of haemoprotein enzymes which represent a large subset of monooxygenases. P-450-mediated monooxygenase activities are almost ubiquitous in virtually all major taxons of living beings, including certain kinds of bacteria and probably all plants and animals. P-450-mediated activities include: hydroxylations of carbon atoms; N-hydroxylations; N-oxidations; S-oxidations; dealkylations; deaminations; dehalogenations; certain azo, nitro, arene oxide and N-hydroxyl reductions. Many of these reactions can produce reactive and toxic intermediates (e.g. Pelkonen and Nebert 1982). The ability of P-450-mediated monooxygenases to generate more toxic products may convert the biodegradation of xenobiotics and pollutants to toxification.

Of significant importance for biotransformations are also epoxide hydrolases, UDP-glucuronosyl-transferases, N-acetyltransferases, glutathion S-transferases, sulfotransferases.

The biotransformation of pollutants may proceed in different components of ecosystems: in the organisms of animals and plants and in "lifeless", "inanimate" components of ecosystems which are substantially influenced by the biota − in water bodies and soils. Such "lifeless", "inanimate" components of ecosystem (biogeocoenosis) have been named by V.I. Vernadsky "biokosny[1] objects" (i.e. lifeless objects that have been formed under the essential influence of living organisms).

2.2.1 Biotransformation of Pollutants in Animals

There are a number of pollutants for which no metabolites have been detected or of which the greater part of the dose is either retained in the body and/or excreted from the body as the unchanged chemical. These xenobiotics are classified as unmetabolized. A balance of more than 95% unchanged (unmetabolized) and less than 5% metabolized is sometimes taken arbitranil to separate unmetabolized chemicals (Renwick 1983). Among the unmetabolized in some mammals and highly environmentally hazardous pollutants are the following: some carboxylic acids (triphenylacetic, phthalic, nitrilotriacetic, ethylenediaminetetraacetic or EDTA etc.), sulphonic acids (benzenesulphonic acid etc.), quaternary amines (tetraethylammonium, Paraquat etc.), chlorinated hydrocarbons (2,4,5,2,4,5-hexachlorobiphenyl, 1,2,4,5-tetrachlorobenzene, 2,3,7,8-TCDD). These xeno-

---

[1] The Russian adjective "biokosny" is difficult to translate with one word. However, it means that the object has properties of both living and lifeless matter (inert, or "kosny" in Russian). This is one of the original concepts of V.I. Vernadsky.

biotics were shown to be unmetabolized in *some* species of mammals. Thus, TCDD is unmetabolized in several species, but in the hamster it is more rapidly excreted as metabolites in the urine (Renwick 1983).

Natural selection in different animal taxa has produced different patterns of biotransformation abilities (e.g. Miyamoto et al. 1988).

One trend in the biotransformation of xenobiotics in animals is the increase in excretability due to increase of polarity, ionization, and hydrophilicity. The main ways to achieve the growth of excretability are: hydroxylation; introduction of ionizable groups, conjugation with water-soluble molecules.

The metabolism of many xenobiotics occurs in the endoplasmic reticulum (the microsomal fraction containing the P-450 cytochrome system) of liver cells. However, some biotransformations are non-microsomal (among them redox reactions involving aldehydes, alcohols, ketons), and other tissues and organs may participate also (lungs, kidneys, skin).

As has been mentioned above, metabolism of some man-made xenobiotics in animals may sometimes give rise to more toxic compounds (Fig. 2.1): in these cases the process is considered as toxification.

Metabolic reactions that may play a role in the toxification of xenobiotics include the following: (1) *oxidations*, among them aliphatic or aromatic C-oxygenations, hydroxylations (Propoxur), metallo-alkane dealkylations (tetraethyl lead), desulfurations (Parathion) etc., (2) *reductions* (azo, nitro, arene oxide reductions etc.), (3) *hydrolyses*, e.g. hydrolyses of epoxides in the case of metabolism of polycyclic aromatics, (4) *conjugations*, including sulphate conjugations of aromatic amines and glutathion conjugations of 1,2-dichloroethane, as well as glucuronidation of some foreign compounds (e.g. see Mulder 1984, Caldwell, Paulson 1984; Miyamoto et al. 1988).

Novel conjugation reactions of xenobiotics with lipids have been described (e.g. Caldwell 1985), which may increase lipophilicity and promote the retention of the compound in the body. Xenobiotic lipids may alter membrane functions.

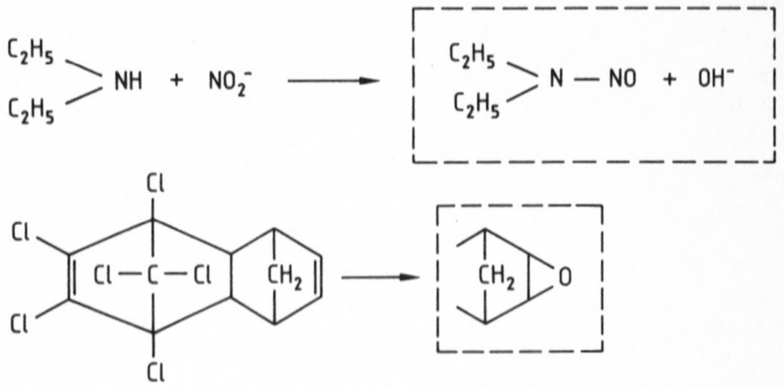

Fig. 2.1. Examples of the toxification reaction of diethylamine (**1**) in mammalian stomach, and pesticide aldrine (**2**) in vertebrate organisms which give the cancerogenetic products (Ostroumov 1986)

Hydrolysis of the acaricide, 2-fluoro-N-methyl N-(1-naphthyl)acetamide (MNFA), yields the toxic product fluoroacetic acid. In guinea pig liver MNFA is hydrolyzed more quickly than in rat or mouse liver, and this may explain the higher toxicity of MNFA for guinea pig than for mice and rats (Paulson 1984).

Another example of toxification is the conjugation of certain arylhydroxy-lamines with N-glucuronic acid and these conjugates in turn are converted to highly reactive arylnitrenium ions under acidic pH which could explain why humans and dogs with acidic urine are more prone to bladder cancer when exposed to arylamines than animals with basic urine (Paulson 1984).

The harmfulness of the products of biotransformation may be mutagenicity, cancerogenicity and teratogenicity.

During the study of the metabolism of organochlorine pollutants, many carcinogenic products of biotransformation have been revealed. Thus, several products of metabolism of DDT are carcinogenic, including the so-called DDE, DD-Cl and DDMU-epoxide. Both the latter are capable of covalent binding with DNA and this may at least in part explain their cancerogenicity for mice.

Additional examples of environmental pollutants that are activated and toxified due to biotransformations include, inter alia,: benzo(a)-pyrene (annual emissions for the US range from 900 t to 1300 t — Dipple 1983), 1-nitropyrene which is activated to reactive electrophile that bound covalently to DNA (Howard et al. 1983), pesticides, e.g. phosphorothionates, Schradan, Methiocarb (Ivie and Bandal 1981), Isofenphos and many others.

The metabolic pathways of DDT and other xenobiotics differ between species; this is the basis for the differingg carcinogenicity of the same compound for different species.

It is important that the molecular transformation of xenobiotics is depen-dent upon the ecology of the given species of animal: an especially crucial role may be played by the microorganisms that inhabit the digestive system of the animal.

Thus in ruminants, the biodegradation of pesticides received with the feed proceeds due to the activity of the microorganisms inhabiting the rumen, one of the parts of the stomach; but drinking water enters the stomach of these animals directly, without passing through the rumen, so that xenobiotics entering the organism of ruminants with drinking water (or through the skin, as is the case with phosphorus-containing pesticides) are especially dangerous. An important practical conclusion is that the milk of ruminants following skin treatment with organophosphorus insecticides may possess high concentrations of poisonous substances.

Some xenobiotics are metabolized through different pathways in vertebrates and in arthropods. On this basis a new class of pesticides is being developed, so called pro-pesticides. Specific proteins, metallothioneins, that are induced in animal organisms by exposure to metals (Zn, Cd, Cu, Ag etc.) may play a role in detoxicating heavy metals in animals. Metallothioneins have been found in terrestrial animals, in fishes and also in invertebrates. However, the environ-mental aspects of their molecular biology are not known in detail.

2.2.2 Biotransformation of Pollutants in Plants

Plants possess no highly rationalized reason to transform xenobiotics to water-soluble form, as animals do for the further excretion of the hydrophilic products. On the other hand, plants effectively exploit binding processes of the xenobiotics or their products and moieties with biomolecules and/or cell walls. In the latter case, molecules of xenobiotics may be "buried" or incapsulated in the polymers of cell walls.

As a result of such bondage, the main structural portions of molecules of pollutants, including pesticides or their toxic residues, may remain unchanged. Hence plants containing such toxic residues may be dangerous as food for man and herbivores. An example is the metabolism of the pesticide cis-anilide in carrots (*Dacus carota*) and cotton (*Gossypium* sp.).

The metabolism of some pollutants in plants may give rise to even more toxic products. Thus, triazine herbicides (Atrazine etc.) are metabolized in the leaves of *Zea mays* into mutagenic products.

It has been shown that plant cells contain enzymes which activate xenobiotics generating mutagenic products in a manner similar to the microsomic fractions from animal liver cells (Gentile et al. 1986).

The biotransformation of pollutants in plants may be disturbed by other anthropogenic factors, with dramatic consequences. When applied alone, the herbicide propanil may control weeds without harming rice. However, when used in combination with certain carbamate and phosphate insecticides which inhibit propanil biotransformation in rice, the rice is also injured. The pesticides Monuron and Diuron are considered to be relatively non-toxic to cotton when applied alone, but are toxic when used in combination with certain insecticides which inhibit their metabolic inactivation in cotton (Paulson 1984). Hence, knowledge of these insufficiently studied issues may help to prevent substantial economic loss.

2.2.3 Biotransformation of Pollutants:
Water, Soils, and Microorganisms

The transformation of pollutants in soils and water bodies is influenced and often even determined by enzymes and metabolites (and products of their chemical and biochemical reactions) produced by different organisms and especially prokaryotes, fungi and algae.

The enzymatic transformation of pollutants may proceed both inside and outside microorganism cells. Organisms may excrete exoenzymes to the environment, where these exoenzymes are able to metabolize xenobiotics or participate in the transformation of dissolved organic matter (DOM) with further consequences for pollution.

Living organisms produce and excrete many organic molecules of ecological importance which act as ecological chemomediators, chemoeffectors and chemoregulators (Ostroumov 1986). It is possible to apply these facts and the concepts developed by V.I. Vernadsky and V.N. Sukachev, about the molecular level of organization of life not only to organisms but also to other ecosystem components including soils and water bodies.

2.2.3.1 Soils

Some molecules of pollutants may be adsorbed on soil particles. This is one of most important phenomena for the fate of pollutants, since the adsorbed molecules are usually beyond the access of soil enzymes. Hence, the adsorption of pollutants on soil particles and colloids may markedly delay the decomposition of their molecules and greatly increase their persistance in the biosphere.

Another phenomenon which may increase the hazard of soil pollution and delay the biotransformation of pollutants is that the latter in some cases inhibit the soil organisms which normally provide the biochemical machinery for the detoxication of xenobiotics.

Chemical or biochemical reactions in soils with the participation of pollutants containing heavy metals or radionuclides do not mean their detoxication for biota, and this generates additional problems in their conservation.

Some pollutants are effectively metabolized and degraded by soil microorganisms only on the intentional addition to the soil of organic substrates ("co-substrates") under conditions of so-called co-metabolism.

Among relatively persistent pesticides which are partly or completely biodegradated in soils preferentially after the addition of co-substrates are the following substances: DDT, 2,3,6-trichlorobenzoic acid, 2,4,6-trichlorophenoxy-acetic acid, Alvison-8 and Ordram (Golovleva and Skrjabin 1979 and others). The molecular mechanisms of connection between the degradation of pollutants and the oxidation of co-substrates are far from being clear.

2.2.3.2 Water Bodies

As in soils, the metabolism of some pollutants in water may require the addition of co-substrates. Thus, the herbicide 2,4-D is destroyed in the water of rice plantations more rapidly if the co-substrate propionate is added (Golovleva and Skrjabin 1979). Hence in natural ecosystems lacking this sort of co-substrate, the biodegradation of such pollutants is slower than in this experiment.

However, the biodegradation of some pollutants in the aquatic environment may be delayed or inhibited in the presence of other pollutants or other organic substances: i.e. a sort of "inverted co-metabolism" or "anti-co-metabolism" may occur. Thus, oil-oxidizing bacteria metabolize oil more slowly when other carbohydrates are present in the medium.

The potential ability of the ecosystems of the Azov Sea for self-purification had been estimated to be equal to a sewage treatment plant which would cost more than $500 million. Man-made changes in the ecosystems of the Azov Sea diminished its annual capacity to detoxify 20,000 t of oil products and 46,000 t of detergents.

The biotransformation of pollutants often leads to an increase in hydrophobicity and this, as quantitative studies of structure-activity relationships show, in turn leads to an increase in toxicity. Thus, enzymatic alkylation of Hg produces more toxic compounds: CH_3HgCl is seven times more toxic than Hg chloride, C_3H_7HgCl is 20 times more toxic, $C_5H_{11}HgCl$ is 300 times more toxic (Florence 1983). The connection between toxicity and hydrophobicity of xenobiotics may be partly explained by the ability of hydrophobic substances to disturb biomembranes and their electrochemical potentials (Ostroumov and Vorobiev 1976, 1978; Ostroumov et al. 1979), and by the high permeability of lipophilic compounds.

Biotransformations of molecules of pollutants in waste waters and sewage sludge are of outstanding importance. Among toxic metabolites that have been found in treated waste water effluents, in activated sewage sludge, in mixed primary and secondary sludge, in aerobically stabilized sludge and especially in anaerobically treated sludge, is 4–nonylphenol (NP), which is the metabolite derived from surfactants alkylphenol polyethoxylates. The concentration of NP in sewage sludge from Switzerland, the FRG, and Finland treatment plants ranged from 0.45 to 2.53 g kg^{-1} (mean, 1.01 g kg^{-1}). NP is more toxic than cadmium for *Daphnia magna* (Giger et al. 1984).

During anaerobic digestion of sludge, no significant biodegradation of surfactants linear alkylbenzenesulphonates (LAS) has occurred. The concentrations of LAS in the sludges represent up to 1.2% of the dry matter of the sludge (MacEvoy and Giger 1985). PCB's, DDT and its metabolite DDE are also found in the sludges: e.g. in Sweden in 1981 the level of PCB's was up to 2.6 mg kg^{-1} and of DDT up to 0.4 mg kg^{-1} (Berglund et al. 1984). Of the 6 million t dry matter of sewage sludge produced annually in the EEC, 30% is used as fertilizer in agriculture, and this leads to the unintentional pollution of the soil by xenobiotics and products of biotransformation of their molecules.

A dangerous process is the interaction of nitrites and amines in sewage waters or in sewage sludge, with the formation of nitrosamines which are potent mutagens and carcinogens.

Among the ecologically hazardous processes which may accompany the biodegradation of molecules of pollutants are the changes in biochemically important parameters of water: levels of oxygen, natural antioxidants and peroxides, pH etc. Thus, during the biotransformation of molecules of the detergent sodium dodecylsulphate (SDS) by *Citrobacter freundii*, the pH of the aquatic medium was shown to decrease (Rotmistrov et al. 1978). Acidification of the water could increase the solubility and toxicity of metals.

Molecular mechanisms of bacterial degradation of pollutants are in many cases an enigma. Thus, many of the biochemical details in the area of catabolism of highly chlorinated phenols remain unknown. Enzymes carrying the dehalogenation and further degradation of pentachlorophenol (annual worldwide production over 50,000 t) and similar molecules have not been characterized (Steiert and Crawford 1985).

The degradation of xenobiotics by microorganisms is retarded by the following characters of molecular structure of the foreign compounds (Cabridenc 1985): (1) aromatic or polycyclical structures, which are broken down more slowly than aliphatic chains, (2) branched chains, which are degraded more slowly than the straight ones, (3) the heteroatoms in a cycle and substitution by halogenated or nitrated radicals, which represent obstacles to degradation, (4) polymers, which generally break down extremely late in comparison with the corresponding monomers.

In conclusion, much more effort should be directed to a more complete study of the molecular mechanisms of action of pollutants on the broad spectrum of organisms (not only the narrow field of those selected for bioassays), as well to the study of the biotransformation of pollutants. The present level of knowledge is dangerously inadequate.

3 Problems at the Ontogenetic Level

Next to the molecular and genetic level of the organization of living matter is the ontogenetic one. At this ontogenetic level individual organisms are considered as the units of life.

Ontogenesis includes the processes of implementing genetic information encoded in the germ cells. During ontogenesis the proper structure (or some portion of a more complicated structure) appears in the proper place at the proper time, the required biochemical reaction or physiological process begins, i.e. the phenomenon of ontogenetic differentiation occurs.

As elementary structures at the ontogenetic level of the organization of living matter we may consider the cells. Man's intervention in the biosphere results in the appearance of multiple environmental factors which are usually detrimental to the normal ontogenetic processes.

3.1 Changes in Embryogenesis

Numerous studies suggest that a great many pollution substances, including pesticides, hormonal agents, metal compounds, hydrocarbons, and other anthropogenic chemical substances, are able to induce disturbances in embryogenesis (i.e. they are embryotoxic) in all tested species of animals or plants such as rats, mice, rabbits, guinea pigs, golden hamsters (*Mesocricetus auratus*), dogs, monkeys, pigs, pike (*Ochotona*) etc. (Fig. 3.1). Substances that induce the development of various malformations are named teratogens (from the ancient Greek "teras", τερας — monster).

When incubating eggs of rainbow trout (*Salmo iridens*) in the presence of benz(a)pyrene at concentrations similar to that of polluted North American rivers, the number of morphological abnormalities in larvae (depigmentation, spine deformation, malformation or absence of eyes, etc.) increased to 14.3% from 2.6% in the control. Similar results were obtained on studying the effect of Malathione on the frog *Microchyla ornata* and in numerous other cases (survey see Mizgireuv et al. 1984). Vast data suggest that the conclusions made above for

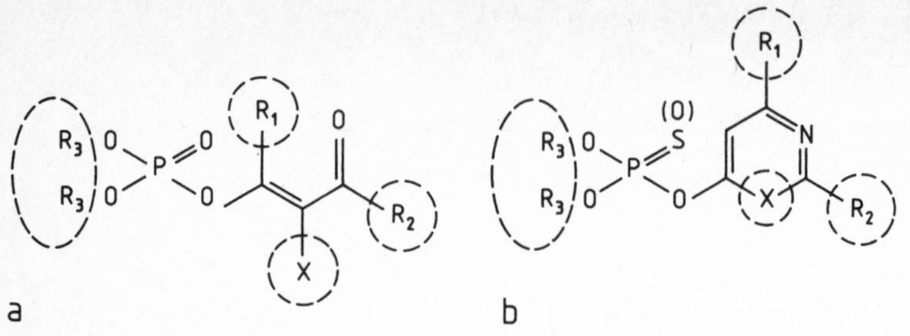

Fig. 3.1a,b. Organic phosphorus pesticides teratogenic for birds. *Dotted* are groups whose specifics determine teratogenic properties. **a** crotonamidphosphates. Substances of the following composition are most teratogenic: $R_1 = CH_3$, $R = NH_2$, $NHCH_3$, $N(CH_3)_2$, $N(C_2H_5)_2$, $R_3 = CH_3$, C_2H_5, $X = H$, Cl. **b** heterocyclical phosphates and phosphorothionates. The following substances of this group are most teratogenic: $R_1 = CH_3$, C_2H_5O, $R_2 + CH_3$, C_2H_5, C_3H_7, $(C_2H_5)_2N$, $R_3 + CH_3$, C_3H_7, C_4H_9, $X = N$, C. (Yablokov and Ostroumov 1985)

experimental situations could be applied also to real situations. The following gives some characteristic examples.

In the 1970's an increased frequency was observed in the appearance of malformed fledglings in colonies of some North American and European tern (*Sterna*) (all these fledglings died afterwards). Analysis showed that reliably larger amounts of polychlorobiphenyls (PCB's) and pesticides were contained in the bodies of the dead fledglings than in the survivors (Culliney 1976).

Avian eggs are sensitive to external exposure to environmental pollutants. Embryotoxic effects have been shown for a variety of pollutants including petroleum constituents, aromatic hydrocarbons, mercury, insecticides, and herbicides. Using externally treated mallard (*Anas platyrhynchos*) eggs, embryotoxic and teratogenic effects have been demonstrated for the insecticides Lindane and Toxaphene and the herbicides Paraquat and 2,4,5-T with formulations and concentrations similar to field applications (Hoffman 1982).

The incubation of quail eggs in air polluted with insecticide DDVP (Dichlorvos, 0.002 mg m^{-3} air) induced both embryotoxic and teratogenic effects. The embryos surviving beyond the 5th day demonstrated multiple malformations, such as lordosis and scoliosis, which may be associated with an anticholineesterase action (Lutz-Ostertag and Bruel 1981).

Many industrial pollutants may also induce embryotoxic effects in avian eggs. Thus, chemicals used in rubber manufacture were embryotoxic for chicken. Among these chemicals are phthalic acid derivatives (cyclohexylthiophtalimide, phtalic anhydride) and three types of oil mixtures known as highly aromatic oils, low aromatic, paraffin base oils and naphtenic oils. The oils and tricresylphosphate were efficient teratogens.

Multiple studies performed in various American, European, and Asian countries have demonstrated that in regions with intense production or use of chemicals, a steady increased rate of congenital malformations and spontaneous abortions is observed (in most cases explained by embryonal development disorders).

PCB's are used as dielectric liquids and also in the manufacture of varnishes, epoxy resins, films, various plasticizers, and additives. Experiments in macaques (*Macaca* spp.) have showed that using feeds containing even low PCB concentration (as low as 2.5 mg kg^{-1}) for 2 months has resulted in severe disorders during pregnancy, foetal resorptions, and abortions, increased neonatal mortality; surviving babies have demonstrated hyperactivity and a decreased ability for learning. In late 1970's the average PCB content in the milk of American mothers was about 1.5 mg kg^{-1} (in some cases as much as 10 mg kg^{-1}), which may have had adverse consequences.

In the 1970's in many countries there was in widespread use a special soap containing 1 to 3% hexachlorophene as an efficient disinfectant (bactericide) and as an acne-treating agent. About 6.1% of the newborn infants of mothers who had used this soap at a rate of 10 to 70 times per day during the whole pregnancy period (some 500 nurses from 6 hospitals in Sweden) had severe genetic disorders, and the total number of newborn infants with congenital defects amounted to 16% in this group.

In addition to chemical effects, embryogenesis may be influenced by various physical environmental factors such as noise, vibration, ionizing or non-ionizing radiation. Thus, in rats, disturbed embryonic development occurred in the following experiment. Pregnant female rats were exposed for 6 h per day (during a total of 20 days) to a non-ionizing microwave radiation with a density of at least μW cm^{-2}; such density values appear to be in the vicinity of major broadcasting, television, or radio stations and radars. Irradiation doses used in medical X-ray diagnosis can adversely affect embryonic development (Sanotsky and Salnikov 1978). Ultrasonic exposure during pregnancy diagnosis can also give rise to an increased rate of congenital abnormalities.

According to WHO, the causes of congenital abnormalities in development are genetic in 25% of the cases and in 5.6% result from extreme environmental factors. A considerable portion of the so-called genetic factors is possibly also due to the mutagenic effect of certain environmental factors (see Chap. 2). It is significant that the number of newborn infants with mental or physical defects increased three- to fivefold from 1965 to 1980 in several countries, including both developing countries (where intense use of agrochemicals takes place) and developed countries.

3.2 Disturbance of Growth

There are numerous data on the growth effect of various anthropogenic factors. Data of both experimental studies and results of changes in populations are presented in Table 3.1.

Table 3.1. Effects of some anthropogenic factors on the growth of some organisms[a]

Factors	Organisms	Effects
γ-Irradiation	Higher plants	Inhibition of growth
Increase of Cu^{2+} in soil	Higher plants	Dwarf forms
Noise pollution of the environment	Higher plants	Delay in leaves and fruit formation
Detergents (including potassium lauryl sulphate, sulfonol and a non-ionogenic surfactant)[a]	*Sinapis alba* *Fagopyrum esculentum*	Inhibition of seedling growth
	Scenedesmus quadri-cauda	Inhibition of culture growth
Pesticide DNOC[a]	*S. alba* *F. esculentum*	Inhibition of seedling growth
Increase of level of O_3, SO_2, NO_x in the air, acid rains	Several tree species	Decrease in growth rate
Heat pollution in the cities	*Pinus sibirica* *Pinus* sp.	Increase in growth
Cu^{2+} and Zn^{2+} in the air and soil	*Pinus silvestris*	Decrease in growth rate, death of small shoots and roots, chlorosis
Air pollution by ore-melting plants	*Pinus silvestris*	Decrease in leaf age
Electromagnetic field (EMF) 50 Hz, 20 kW m^{-1}, 168 h	*Picea abies*	Decrease in growth rate
The same, 24 h		Elongation of roots
Organochlorines	Some species of marine phytoplankton	Decrease and end of growth
Toxins of cyanobacteria (due to euthrophication)	Some species of fishes	Decrease of growth
Some specimens of crude oil, per os, 0.2 ml and more	*Larus argentatus*	Decrease of growth
EMF	White mouse	Decrease of growth

[a]Experimental data by S.A. Ostroumov, V.N. Maximov, H. Nagel, S.V. Goryunova, T.N. Kovaleva et al. (Ostroumov et al. 1988; Maximov et al. 1988a, 1988b).

Insect growth-disturbing agents are of increasingly widespread use in plant protection. Such agents (when exhibiting a non-specific action) can influence the growth and development of other Arthropoda species which are not the "targets".

The growth of many plant species has been shown to be sensitive to SO_2, NO_x, and O_3. Among them is timothy grass (*Phleum pratense* L.); its net assimilation rate was reduced by an average of 14.7% by 0.12 ppm SO_2 and the sensitivity of this species to SO_2 was affected by photon flux densities and temperature (Jones and Mansfield 1982).

In the USA, the economic loss due to decrease in growth of plants which is induced by pollutants has been estimated at about US\$ 2×10^9 annually.

The exposure to pollutants, including petroleum, has been shown to inhibit the growth of several sea hydrobionts. Thus, the growth of sea star embryos has been inhibited by petroleum and no long-term adaptation to petroleum has been detected in studies in the Lawrence Livermore Laboratory, University of California.

Thus, very different anthropogenic factors may frequently cause disturbance of the growth of organisms.

3.3 Disturbance of Reproduction

Disturbances of sexual maturation and reproduction are frequently observed as a result of anthropogenic action.

Thus, faulty timing in developing gonages of the male and female frog (*Rana ridibunda*) has been found due to the accumulation of several toxicants (mainly pesticides) in the bodies of animals in the North Caucasus and the Lower Volga River region, and this has excluded the ability to propagate successfully (Koubantsev and Zhoukova 1982).

Thinning-down of the egg shells of many predatory birds became one of consequences of the widespread use of DDT in the 1960's. Thus, for example, only four feathered fledglings successfully hatched from 1125 nests in a brown pelican (*Pelicanus occidentalis*) colony at Anacanapa Island near Los Angeles, California, during the 1969 season. Such a drastic egg loss has been associated with the breaking of the thin-shelled eggs by the nesting birds. The DDT/DDE concentration in ruptured eggs was about 1200 ppm, whereas that in unruptured eggs was about 900 ppm. The DDT/DDE concentration in the anchovy, principal food of the pelicans, amounted to about 4.3 ppm in 1969. When the DDT content dropped in 1974 in anchovy (*Engraulis*) 28-fold and in the pelican eggs ninefold, one fledgling developed successfully in every nest (Culliney 1976). In the early 1970's the white-tailed sea eagles (*Haliaeetus albicilla*) nesting along the Finnish shore of the Botnic Gulf had an egg shell thinner by 16 to 19.7% than before the year 1935, and the fledglings hatched from only two out of ten eggs laid (Fig. 3.2). In the late 1970's, when DDT use was strictly restricted, its concentration in the eggs lowered, the egg shell thickness increased and nesting success rose (Fig. 3.3) (Koivussari et al. 1980).

The PCB's have a destructive effect on reproduction in bird species. In the female Japanese quail *Coturnix coturnix japonica*, the PCB Clophen A60 caused delayed laying and a diminished laying capacity, and the breaking strength of the eggs was reduced. In males, PCB treatment resulted in a trend towards reduced testis weights (Biessmann 1982).

Organochlorine compounds cause severe reproduction disorders in mammals. The DDT/DDE concentration in the California sea lion (*Zalophus californians*) living on the Channel Islands near Los Angeles amounted to hundreds of ppm in the late 1960's. As a result, cases of premature birth became more frequent and the newborn death rate increased. The pesticide concentration in the bodies of the prematurely parous female animals was two to eight times

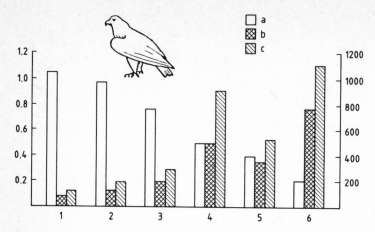

Fig. 3.2. Average annual productivity (the number of young birds per pair of adults) (**a**) and the average level of DDT/DDE (**b**) and PCB (**c**) in the fatty tissue of white-tailed eagles from various habitats. *1* Greenland; *2* North Norway; *3* Swedish Laplandia; *4* West Germany; *5* Finnish coast of the Gulf of Bothnia; *6* Middle Sweden. (Helander et al. 1982)

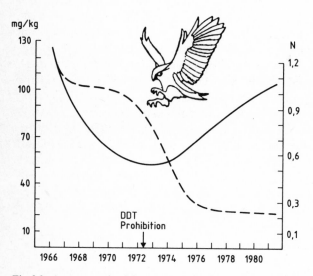

Fig. 3.3. Average productivity (number of young birds per pair of adults) and DDT/DDE content in the eggs of bold eagles in the north-western part of Ontario (Canada). It shows that following the ban on DDT in 1972 the DDT/DDE content dropped sharply and the number of nestlings began to increase

higher than in the bodies of female animals with a normal matured foetus (Culliney 1976). In the Gulf of Bothnia, in the Baltic Sea, only 27% of the female grey seals (*Halichoerus grypus*) were pregnant in the mid-1970's, whereas in other oceanic regions with lower PCB content, up to 90% of the female animals are usually pregnant. Only one of ten pregnant female animals was able to give birth to normal calves, which was associated with pathological uterus disorders (Olsson 1977, and other works). Our studies have shown significant spermatogenesis disturbances in male animals from this region. An ever-increasing number of similar data related to humans are in good conformity with these findings. Numerous studies had demonstrated an inverse dependence between the concentration of organochlorine compounds (PCB's, in particular), Pb or Cd compounds and sperm quality: spermatozoid mobility and concentration in ejaculates have been significantly disturbed, the number of pathological spermatozoids has increased (Stachel et al. 1983, and other works). Noxious effects of PCB's, DDT, and other organochlorine compounds on the course of pregnancy have been indicated also in many other studies.

Pollutants with oestrogenic properties may disturb normal reproduction processes in wild animal populations. Among the pollutants with various degrees of oestorgenecity are (Rall and MacLachlan 1980): diethylstilbestrol (DES, annual use in the USA is over 27,000 kg); o',-p-DDT (an isomer which accounts for 15 to 20% of the commercial mixture of DDT); PCB's, especially Arochlor 1221; PAH, and possibly some others.

Small doses of pesticide intensify reproduction in some cases. Thus, black ducks (*Anas rubripes*) receiving up to 50 µg/kg of the insecticide Toxaphene (polychlorocamphene) in food for 19 months have exhibited a lowering in the number of days needed for completing the clutch and an increased ability to propagate in the second year of the experiment (Haseltine et al. 1980). Similar findings are known for insects.

Crude oil spill pollution is detrimental to reproduction in multiple species of animals and plants (survey see Mironov 1985).

Other anthropogenic factors also influence reproduction. In one of a series of experiments, mice were "sonicated" for 2 h daily for a total of 1 year with noise recorded in the New York underground. As a result, almost all the male animals lost the ability to fertilize the females. The female animals sonicated either became totally unable to bear or had a significantly decreased number of young in the litters. The effect of sound waves generated by modern high-speed aircraft and space ships is shown to be noxious for reproduction. In particular, low frequency noise which is generated at the sonic barrier transition by supersonic aircraft has a negative effect on clutch size and fledgling hatchability in colonies of herring gulls (*Larus argentatus*). The action of the low frequency noise generated by the US space shuttle ship *Columbia* has also proved dangerous; embryo implantation has been delayed and the number of spontaneous abortions has increased in the California sea lion (*Zalophus californianus*) colony inhabiting the shore within the noise-affected area near the Californian shore (W. Evans, pers. commun.).

Noise pollution (65 to 68 dB, 12 h) has been shown to decrease the concentration and mobility of spermatozoids in rabbits; the efficiency of fertilization has been reduced.

Exposing rats to an electromagnetic field (EMF) having an intensity of 1 to 10 μW cm^{-2} (wavelength 12 cm) has resulted in a lowered number of successful copulations, the appearance of less viable descendants and reduced litter size. Chronic action of the 10 to 50 μW cm^{-2} EMF (wavelength 10 to 12.6 cm) as well as an ultrashort wave (wavelength ca. 1.9 m) has given rise to oestral cycle disorders in mice, an increased number of sterile animals, a significant lowering of the number of newborn, and an increased number of still births (Nikitin and Novikov 1980). The 5 kHz EMF (a field strength of at least 4 kV m^{-1}) has exhibited a gonadotoxic action on rats (Menshikova 1983).

Many birds in nature abandon their clutches and discontinue reproduction when they are scared off nests several times. To cause Asiatic white crane (*Grus leucogeranus*), one of the world's rarest cranes, to abandon sitting on eggs it is enough for a human to appear only once within a radius of several hundreds meters from the nest.

Up to recently, changes in the ratio of the sexes had not gained attention as a possible consequence of pollution. However, in Central Scotland, in a steel mill neighbourhood, a reliable shift in the sex relationship between newborn children has been found: within the highly polluted area the number of newborn boys was definitely larger (by 37%) than the number of newborn girls (Lloyd et al. 1985).

The situation with the grey seal (*Halichoerus grypus*) in the Baltic Sea, whose numbers are continuously diminishing despite the commercial catch having been stopped and protection in numerous reservations, and which seemingly can disappear in the Baltic Sea completely in the near future, could be a serious lesson. The same fate may await all long-living animals that are able to accumulate pollutants in dangerous concentrations in their bodies.

3.4 Metabolic Disturbances, Poisoning and Diseases

This section considers briefly some problems of disease origin in the ontogenesis of the mature organism as a result of the action of anthropogenic factors. There are numerous publications on this subject and there is no need to consider this problem in detail here: it is only important to show the position of these postnatal ontogenetic disturbances in the general picture of anthropogenic effect at the ontogenetic level.

For convenience, the material of this section is organized into subsections devoted to the pollution of individual components of the environment. Such a division is to a great extent conditional (e.g., it is difficult to classify pesticide pollution according to components of the environment, because the pesticides may initially enter soils through the air and can then be carried out into the world oceans, etc.).

3.4.1 Air Pollution

Air pollution contributes to extreme metabolic disturbances or various diseases in plants. Evident damage to plants is found under the action of acid rain at pH 2 to 3.6. Low pH inhibits the spermatozoid mobility in gametophytes of the fern *Pteridium aquilinum* and reduces fertilization success. Motor vehicle emissions and other municipal pollutants are also very detrimental to plant physiology. Thus, in cases of English oak (*Quercus robur*), small-leaved lime (*Tilia cordata*) and elm (*Ulmus laevis*), decreasing chloroplast amounts were found, reducing the number and dimensions of leaves, shortening the leaf life-time by 25 to 40 days, reducing size and density of stomata, and reducing total chlorophyll content 1.5 to 2 times.

SO_2 and NO_x pollution of air causes abrupt changes in the frost sensitivity of certain herbaceous plants; fluorine-containing industrial emissions appear to be the cause of various disturbances in almost one half of the plants in the oak and hornbeam community studied (Goryshina 1982).

Under the influence of the EMF malformed organs and structures can be developed in plants.

Many pesticides (Carbophos, Metathione, hexachlorocyclohexane, etc.) are responsible for haematological index shifts, and changes in blood serum proteins. There are numerous examples of how very "common" atmospheric pollutants affect animal ontogenesis.

For 104 regions in Great Britain, examining comparative data on the total mortality rate for the years 1960, 1969, and 1974, a reliable correlation between total mortality rate and atmospheric sulphate content has been revealed. A similar analysis of data for London seems to show a reliable correlation between mortality rate and particulate content, smog level, and SO_2 concentration (Ostro 1984). A reliable correlation was demonstrated between total sickness rate and average annual concentrations of industrial air pollutants, even in cases when the concentrations of the individual pollutants had not exceeded the admitted values (Tabakova 1985).

Every year during the 1970's, in the developing countries alone, some 375,000 persons were poisoned with pesticides, and of these some 10,000 died. Statistical data suggest that within certain regions of the USA (e.g. in California) agriculture has become one of the most dangerous occupations according to the mortality rate caused by occupational risks: most cases are associated with the direct poisoning with pesticide or secondary diseases resulting therefrom.

Tests in monkeys and observations in humans show that one single exposure to some organic phosphorous pesticide of the Dieldrine, Malathione, and Parathione type contributes to an encephalogram change, i.e. changes in brain activity, such encephalogram changes remaining for about 1 year after exposure. In humans such exposures results in disturbances of sleep or memory, loss of libido, increased irritability, and difficulty in concentration.

Particularly dangerous is the pollution of the atmosphere with certain anthropogenic aerosols (e.g. based on organofluorine compounds) that results in damage to the ozone layer in the upper atmosphere. Diminishing the ozonosphere depth is directly associated with an increase in ultraviolet radiation intensity. While the effect of a deterioration of 1% in the ozone shield above mainland USA for 1980 seems to be twice as low on the average as that predicted in the 1970's, nevertheless according to some authors it has resulted in an increase of 2% in skin neoplasm incidence throughout this region (Faber 1982). One of the most dangerous pollutions for humans (and many other mammals or vertebrates in general) appears to be air pollution with PAH's, and common benz(a)pyrene among them. An exposure to PAH's contributes to various neoplasm developments, including the formation of malignant tumours.

An interrelation between air pollution level and lung cancer incidence rate has been established for many countries (including the USA, Japan and the USSR). An analysis of lung cancer mortality rate in areas of the USA where mining and melting of lead and zinc were performed in the states of Oklahoma, Kansas and Missouri up to 1960 has demonstrated that its values substantially exceeded the average national level for both the 1950 to 1969 period and for 1973 to 1977. This might be associated only with consequences of the chemical pollution of air before 1960. Among 85 dyes of pigments studied, 25 were found to be carcinogenic and 14 are suspect in this respect (Grushko and Timofeeva 1983). In some cases, surprisingly low concentrations appear to be carcinogenic. Thus, a dose of 100 ng kg^{-1} body weight of 2,3,7,8-tetrachlorodibenz-p-dioxine has induced cancer tumours in 50% of animals in chronic tests (Gold et al. 1984). Artificial EMF of even low or moderate intensity can produce an adverse effect on the state of a vertebrate organism, changing the course of the oxidation-reduction processes and the oxygen condition in tissues, inducing functional changes in the cardiovascular or endocrine system. About 1% population in the USA is continuously subjected to an EMF exposure exceeding the 1 μW cm^{-2} level (Fig. 3.4). Exposing rats to the 500 μW cm^{-2} EMF for 7 h every day for 3 months resulted, among other changes in physiology and behaviour, in decreased content of the thiol groups and cholinesterase in the blood, an increased shock excitation threshold, and a changed α-rhythm in bioelectric brain activity. A similar irradiation of pregnant rats for 20 days resulted in depressing the behaviour of newborn rats. The change in the state of the nervous system in rats was also observed at the lower field intensity (the energy flux density of at least 10 μW cm^{-2} for the exposure time of 2 h per day); such an energy flux is equivalent to that induced in the vicinity of radio transmitter, television broadcasting, and radar stations. It has been noted that the youngest and the oldest individual animals have exhibited particularly high susceptibility. Exposure to low intensity microwaves (of at least 1 μW cm^{-2}) causes disturbance in behaviour, emotional state, and in the autoallergic process (Shandala et al. 1985).

Not only one single EMF irradiation, but also the cumulative effect can be dangerous. There are data on the detrimental consequences of irradiation at small intensity. Rabbits and guinea pigs have developed leukaemia and changes in the immune and endocrine systems under such conditions. Irradiating mice and rats with small intensity has resulted in evident behaviour modification and a rise in body temperature (that has been confirmed in monkeys), and such an irradiation

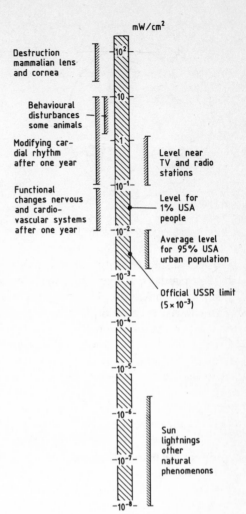

Fig. 3.4. Influence of microwave radiation on living organisms. (Based on data from various sources in Yablokov and Ostroumov 1985)

has been shown to be very probably carcinogenic and cocarcinogenic. While there are at present no reliable data on the effect of ultra-low frequencies generated by electrified vehicles and industrial installations, it is known that some of these frequencies coincide with those of certain rhythms in mammal brain activity. In some experiments, the modulated ultra-low-frequency EMF has affected ion transport in the test animal's cells. The range for the possible action of such anthropogenic fields is very significant. There are growing data on the detrimental effect of electric power transmission lines on animals (survey see Protasov 1982; Fig. 3.5).

The last examples in this series concern radioactive pollution. Experimental studies showed that the tree swallow (*Iridoprocne bicolor*) and house wren (*Troglodites aëdon*) either avoided man-made wooden boxes for nesting which

Fig. 3.5. Stages in hydra degeneration under the impact of weak direct current whose intensity is similar to that of current induced by power transmission lines in natural habitats. (Protasov 1982)

had been suspended within an increased radiation zone (γ-radiation of at least 35 μCi kg^{-1} per day), or abandoned the nests before coming to lay (Zach and Mayoh 1982). It is known that lichens exhibit an ability to accumulate radionuclides (particularly, because of their significant life span). The malignant neoplasm mortality rate for the aboriginal population of some Arctic regions appear to be higher than average. Researchers associate this fact with the sharply increased radionuclide intake into the human body in the North along the following food chain: lichen — reindeer — humans (Troitskaya et al. 1982).

3.4.2 Hydrosphere Pollution

A substantial part of the pollutants enter the world oceans not via river drainage, but via rain and atmospheric dust. The role of pollutants entering the oceans directly via freshwater discharge is also great. Millions of fish have been killed in the lower reaches of the Mississippi River as a result of putting Endrine (one of the most potent pesticide) into the water at the beginning of the 1960's. Cases of fish poisoning with pesticides are known for the estuaries of several Central American rivers, as well as for rivers in the Cameroun, Zimbabwe, and some other African countries.

Even trace amounts of DDT or Chlorophos have produced respiratory depression, and avitaminosis in Gammaridea invertebrates and some species of fish. Certain pesticides in extremely low amounts markedly affect the behaviour of fish and invertebrates. Thus, Sevin and its derivatives in concentrations of 10^{-4} ppm disrupt the gregarious behaviour of the fish *Menidia* (Culliney 1976).

In tests using the fish *Micropterus salmoides* as an indicator, it was possible to record minimal concentrations of Hg, Cu and cyanide at a concentration of 5×10^{-2} ppm, ammonium (in terms of nitrogen) at 1 ppm, and Cd at 0.2 ppm. Trout is susceptible to some water-soluble substances at concentrations of about 10^{-5} ppm.

PCB's, pesticides and other pollutants affect bird behaviour: birds demonstrated a slowed avoidance of danger and absence of fear of humans. Usually such birds exhibited a state of exhaustion. Many organochlorine compounds are lipophilic and readily soluble in fats (lipids). When the bird is well fed, most such toxicants are concentrated in the relatively inert fatty tissue. However, when it is necessary to spend this accumulated fat in emergency, they poison the organism. Such cases of poisoning are known not only for birds at the end of the transmigration, but also for bats upon awakening from hibernation (Clark and Krynitsky 1983).

Signs of fin destruction have been revealed in 30 species of bottom-dwelling fishes along the Pacific coast of the USA as a result of discharging industrial and municipal waste effluents from this densely populated region directly into the Ocean through pipelines flowing offshore (Culliney 1976).

Methylmercury chloride (MMC) (in drinking water solution) induced changes in sleep-waking rhythms in rats. Sleep-waking disorder was shown to appear at lower concentrations of Hg in the brain and at a shorter latency than the behavioural disorders of movement and postural maintenance previously reported. MMC produced an increase in both dark-phase slow wave sleep and paradoxical sleep (PS) as well as a decrease in light-phase PS at the expense of a growth in light-phase wakefulness and a delayed phase of the circadian PS rhythm. Insomnia and hypersomnia were observed due to contamination by organic Hg in Japan, Iraq, Ghana, and in other countries among workers dealing with alkylmercury compounds.

It has been shown that the fish cancer incidence rate within at least five major drainage basins of the USA is dangerously significant: in the Hudson Bay up to 80% Atlantic tomcod (*Microgadus tomcod*) were affected with liver cancer; about 80% gobies of more than 3 years age in the Black River (Ohio) were affected with skin or liver cancer; Canadian pike perches *Stizostedion canadiensis* and *S. vitreum* inhabiting Torch Lake, (Michigan), poisoned with local copper mine wastes, were completely affected with liver cancer, ossificating fibroma and spleen lesions. In the control tests, a deposit sampled from Buffalo River (New York) induced skin cancer in gobies after 1 year. Moreover, when chemical recovered from this deposit were added to the goby feed, eight fishes of ten started to suffer from severe liver lesions, including cancer. The incidence of malignant neoplasms has been noted also in the Baltic basin in pike (*Esox luceus*) and trout (*Salmo trutta*) (Bogovski et al. 1982).

One of the most widely used insecticides is Toxaphene, a complex blend of chlorine-containing organic compounds. For more than a decade (from 1964 to 1975) it was used in the USA and then in Western Europe more intensely than any other composition, at the rate of about 150 g/year per habitant. As a result, its concentration in the atmosphere over the World Ocean exceeds that of other pesticides. This pesticide, even at a negligible concentration of 8×10^{-3} ppm, appears to kill some fish (gambusies, *Gambusia affinis*), and induce irreversible changes in liver, lungs, gills, spine of freshwater catfish (*Silurus*), trout (*Salmo trutta*), pimephales (*Pimephales*), and other fishes (Culliney 1976).

River waters appear to be polluted with surfactants which can affect terrestrial plants on irrigated lands. Experiments by S.A. Ostroumov and V.N. Maxinov in 1985–1986 showed that anionic or non-ionic surfactants can disturb the normal differentiation of rhizodermic cells and root hair formation in *Sinapis alba*, *Fagopyrum esculentum* and *Triticum aestivum* (Ostroumov, Maximov 1988).

Detergents were shown to inhibit the elongation of plant seedlings in several species and inhibit the growth of algae cultures (Goryunova and Ostroumov 1986; Maximov et al. 1986).

PAH's entering the sea with petroleum or by some other route appear to induce tumour formation in fishes or amphibians. In the USA, over a period of 6 years, one distinct population of amphibian *Ambystoma tigrinum* inhabiting a lagoon heavily polluted with PAH's was studied. From 30% to 50% of all individual animals observed (a total of about 28,000 animals were studied) had neoplasms or cysts of various types in the skin. Comparative examination of 16 other populations of the same species revealed no such skin neoplasms in the animals (Rose and Harshbarger 1977).

Some comparative studies have been carried out in the USA and Canada for the last 5 years to establish a relationship between aquatic reservoir pollution level and sickness rate in fish. Within the most polluted aquatic reservoir (about 3 to 50 mg organic substances per 1 l water) it has been found that more than 8.5% fishes suffered from diseases of bacterial, fungal, or parasitic nature, and more than 8% had spinal disorders. Within the low polluted aquatic reservoir sick fish fraction was only about 1%, while only about 3% of the fish had spinal disorders.

The petroleum contamination of the world oceans is significant. Many algae, invertebrates, or fish die at a crude oil concentration of 1 to 50 ppm (Mironov 1985; etc.). The consequences of sea contamination with crude oil are dramatic for birds and mammals also. Bird feathers, normally non-wettable with water become impregnated with crude oil, the body temperature cools rapidly, and the birds perish. Seals, *Sirenia*, and *Cetacea* perish as a result either of petroleum directly entering their organisms, or of eating poisoned fish, invertebrates and algae. A recent survey (Meyers and Hendricks 1982) reports numerous data on pathological changes in organs and tissues of various fish and aquatic invertebrate species induced under exposure to various pollutants. A description of the effect of chemical pollution on physiology, behaviour, ecology and distribution of marine or limnetic inhabitants within natural populations is the subject of numerous volumes and many scientific conferences are devoted to this problem.

The effects of thermal pollution on aquatic inhabitants are very important. As a result of major nuclear power plant operation, significant areas of warmed water are produced that are higher in temperature by several degrees than the average value for the surrounding water area. These warm water areas attract many fish sensitive to the water temperature gradient across tens of metres in a horizontal plane. Fish remains more active in warm water during the cold season, growth being more intense. These seemingly more comfortable areas, however, become fatal traps when the temperature of the inflowing water is abruptly varied. Thus, when the Oyster Creek nuclear power plant (New Jersey, USA) was closed to refuel for several hours in January 1972, the discharge area temperature dropped from 15°C to almost 0°C. Tens of thousand of fish crowded here for wintering perished, since they failed to survive such an abrupt temperature change. Moreover, when the temperature is varied, an oversaturation of water with dissolved air may occur that results in embolism in the fish (blood vessels plugged with nitrogen bubbles), leading to death.

Even sublethal, and sometimes also hardly experimentally detectable exposure are dangerous in natural populations. Shrimps containing traces of Arochlor-1254 (one of the common polychlorobiphenyls widely used in industry, which is one of the components of global biosphere pollution in their bodies were transferred from sea water, which is habitual for them, into brackish water. The sea salt content in the latter was lowered fivefold in comparison with sea water. All the shrimps perished, since the presence of PCB had inhibited the saline concentration in the blood, whereas the control group with no PCB traces survived. Similar desalting of sea water near the shore often occurs after heavy rains in habitats of the shrimp *Penaeus* (Culliney 1981).

Sometimes the influence of the pollutant affects seemingly minor biological features. Under experimental conditions, the brackish-water bivalve Lamellibranchata *Rangia cuneata* was fed with insecticide-contaminated phytoplankton. No pathological changes in the organism were evident, except for disturbances in the secretion of byssus, the filament used for attaching the mussel to the substrate. It is evident, however, that mussels not attached to the substrate are threatened with immediate death in nature.

Variations in the behaviour of aquatic organisms have been revealed under exposure to various physical fields of an anthropogenic nature. Thus, it was found that using electric trawl nets resulted in premature spawning of some fish species. Electric fields which are generated by the high voltage power transmission lines affect fish distribution (even to ceasing migration), and the low frequency hydromechanical fields which are produced by running ships appear to compel many fishes to avoid maritime routes (Fig. 3.6).

3.4.3 Soil Pollution

The presence of Mo, Cu, Zn, and other metal compounds at unusually high concentrations in soils may result in damage to various organs or tissues in plants, development of pathological neoplasms, disturbance in the development of morphologic structure. Applying salt for cleaning streets or highways from snow results in leaf necrosis in many woody varieties.

Poisoning fowl with lead shots is becoming widespread in Western Europe and the USA. Ducks swallow shot as gastrolites, i.e. pebbles helping to grind the food in the stomach. Six lead shots of medium size swallowed by the mallard *Anas platyrhynchos* suffice to cause lethal poisoning with Pb, while sublethal doses of Pb affect reproduction.

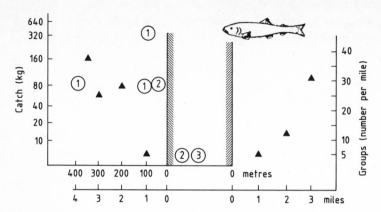

Fig. 3.6. Distribution of fish runs relative to shipping routes in the Votkinskoye Reservoir (in metres) and the Black Sea (in miles). Big fish runs and sprat shoals keep away from the fairway. *1* common bream (*Abramis brama*); *2* roach (*Rutilus rutilus*); *3* ruff (*Acerina cernua*); the *black triangles* are sprat (*Sprattus sprattus*) shoals. (Protasov 1982)

Using the analysis of soil samples as a basis, it has been established that the high average content of compounds of Mg, Pb, and Co in soils are related to the overall rise in the incidence of recorded oncological diseases.

Increased concentration of Pb in the human organism has a negative influence on mental activity (especially in children). Studies performed independently and using different materials by researchers from the FRG, the USA, and England have showed that children with increased content of Pb in the blood (more than 13 μg per 100 ml), were constantly and statistically reliably assessed with a lower score in the well-known IQ test, the influence of the social and economic status of their families having been carefully excluded (Ratcliffe 1983). Level of Cd or Pb in the hair of mentally retarded children were found to be higher than for the group of normal children (e.g. Jule et al. 1982).

3.4.4 Pollution of Animal Feed

Feed of animals may be contaminated by pesticides, PCB's, and other organic and inorganic pollutants.

In the mid-1960's, US mink ranchers experienced a decline in reproduction in minks that was eventually traced to the PCB content of Great Lakes fish which was part of their diet. Minks fed PCB's at 5 ppm have complete failure in reproduction. Mammals fed hexachlorobenzene (HCB), which is used in agriculture and industry, demonstrated hepatomegaly, porphyria, and immunosuppression (Rush et al. 1983; Bleavins et al. 1982, etc.).

The toxic effects of different heavy metals have been documented using many species of animals. Aluminium compounds as dietary supplement also lead to a diversity of toxic effects in rats, rabbits, lambs, and other organisms.

Many pollutants change the normal forms of behaviour of animals. Such changes have been described in fish, mammals and birds (for instance in black ducks, *Anas rubripes*), where a diet containing low-level Cd had induced altered avoidance behaviour in response to flight stimulus.

Many pollutants which contaminate feed may induce carcinogenesis. Carcinogeneicity has been shown for benzene administered with corn oil to rats and mice and for other pollutants.

Contamination of plants which are used as feed by animals seems to be a serious hazard. Unfortunately, the toxic effects associated with this manner of action of pollutants have not been investigated extensively.

The importance of this manner of action of pollutants in animals is especially high, due to the well-demonstrated bioconcentrations of many pollutants in the organisms which are the food resource for the animals of the next trophic level.

3.4.5 Factor Interaction

In nature, the action of any individual factor is generally complicated, specifically amplified or weakened by actions of many other factors. The following examples serve to illustrate this.

SO_2-contaminated air at a concentration of 10.0 to 1 mg l^{-1} (an exposure of up 21 days) has been reliably reported to increase mice mortality caused by bacterial infection. Chloride accumulation has been positively correlated with leaf necrosis in horse chestnut (*Aesculus hippocastanum*) on the outskirts of Bern, Switzerland. However, no such a correlation between leaf necrosis and chloride content was revealed in the central and industrial areas of the city with the most intense CO and NO_2 air pollution. This paradoxical result is explained by an inhibition of the stomatal leaf apparatus by atmospheric toxicants.

The effect of Cu on the plant organism is sharply amplified in the presence of Pb salts. Copper also increases radioactive effects on plants, whereas the presence of Ba, Mn or Mg salts weakens such an effect.

When studying the effect of Cu, Co, Cd or Zn compounds on the shrimp *Calianassa australiensis*, it was found that any two-metal combinations are more toxic than a single metal, and any three-metal combinations are more toxic than the effect of any compounds of two metals.

The natural resistance of albino rats to the plague was lost even under the influence of small doses of the anticoagulant Varfarine.

About 70% of cardiovascular and pulmonary diseases of women in urban areas are associated with the combined action of chemical contamination and noise. A similar effect is produced by the combined action of noise and air pollution on children: a noise of 40 to 60 dB in combination with the action of carbon monoxide in a 1 to 2 mg m^{-3} concentration (i.e. the situation characteristic of streets in large cities) affects the health of preschool children.

Penetrating neutron radiation is in some manner able to damage in the dog the natural system of resistance to Siberian plague. As for diphtheric toxin, the effects of short-wave UV-radiation and long-wave UV-radiation on rabbits differ highly: the former increases their resistance, while the latter decreases it.

As is well known, UV radiation that in natural conditions causes the generation of the most toxic smog forms (as a result of photochemical reactions involving sulphur and nitrogen compounds). Exactly this photochemical smog developed in Central Europe under the particularly abundant sunshine in 1976, that resulted in the needle damage over vast forest areas in the Schwarzwald (FRG) in spite of a relatively low level of air pollution in this region.

A group of substances has been picked out which become evident carcinogens only in the presence of some other agents (so-called co-carcinogens). Many emulsifiers (such as Tweens, Spans), phenols, the antibiotic Griseofulvin, dimethyl sulphoxide, the herbicide 2,4-D, and other substances have been proved to be co-carcinigens of this type.

Finally, it must be said that hopes of controlling sea oil spills by using dispersing agents do not seem to be coming true, since petroleum in combination even with the most low-toxic dispersing agents proves to be much more toxic for fishes, crabs, and other aquatic organisms.

3.4.6 Non-Specific Action of Pollutants

The influence of pollution manifests itself rather in a general weakening of the organism, lowering the level of resistance to other environmental factors which are not directly associated with the effect of specific pollutants, than by affecting individual organs or systems. It magnifies the duration, frequency, and severity of common diseases.

The availability of nutrients may be varied and the susceptibility to different diseases may be amplified as a result of metabolic disturbances caused by pollutants. Data of this type are being accumulated for animals, plants and humans.

The study of several thousands of birds and bats which had died from various causes in Lower Saxony and other neighbouring districts of the FRG revealed that species containing significant amounts of pesticides, heavy metal salts, and other chemical pollutants (assays were made for 50 different toxicants) in their bodies had more frequently died from "natural" diseases and had fallen prey to a predator than non-polluted species.

People living in the vicinity of chemical plants (and not working in these plants) have had colds 5 to 10 times longer than in the control. Diseases of the liver, kidneys, and pancreas have been found to be from 1.5 to 2.5 times more frequent in the test group of individuals than in the control group. The overall sickness rate for preschool children living in conditions of polluted air have been found to be 30% higher than for the control group. A higher sickness rate has been found to be characteristic for children up to 15 years age living in areas with an intense use of pesticides, the proportion of pesticide action on the larger part of nosologic forms being from 14.1% to 56.6%.

Overall non-specific resistance to bacterial infection may be lowered in mammals under the influence of heavy metals.

Even very low doses of 2,3,7,8-tetrachlorodibenzofurane (TCDF) and tetrachlorodibenzodioxine (TCDD) appear to act as immunodepressants in mice (Josephson 1983). TCDD is a contaminant in some herbicides, and both TCDD and TCDF could be formed when burning organic chemical substances.

Among the pollutants which have immunomodulatory effects on animals are also PCB's, polybrominated biphenyls (PBB's), DDT, HCB, 2,3,7,8-TCDD, and others. Hexachlorobenzene (HCB), which has been detected in fish, in the body fat of domestic animals and poultry, and in human milk, has been reported to cause humoral immunosuppression in mice and to depress lymphocyte responsiveness in mink (*Mustela vison*) and in European ferrets (*Mustela putorius furo*). HCB (25 ppm dietary, 8 months) depressed the lymphocyte blastogenic response to lectin. The in utero and early postnatal exposure of young mink to HCB also decreased lymphocyte responsiveness (Bleavins et al. 1982).

In rabbits, the level of lysozyme, which is an important bacteriocidic agent participating in resistance against infections, decreased after inhalation of low Ni concentration 0.1 mg Ni m^{-3}. For comparison, in the USA, the threshold limit value for Ni is 1 mg Ni m^{-3}. The Ni-induced level of lysozyme in lung lavage fluid was lower than 0.04 μg ml^{-1} versus 2.3 μg ml^{-1} in the control (Lundborg and Camner 1982).

The ability to suppress the human immune system (immunosuppressive action) has been shown for a number of herbicides such as Cotorene, Monuron and Dosanex (Methoxurone).

The action of many of the factors of environmental pollution appears to be different for differing species (Hemminki 1981; Murty et al. 1983; etc.). These differences usually concern the susceptibility of various species to a particular factor. Thus, fish are more sensible to some pesticides (e.g. Phenotrothione) than mammals. Studies of the action of several chemical compounds has shown that rats and mice seems to be less sensitive to environmental pollution by these than, for example, hens, rabbits, monkeys or humans. S.V. Goryunova and S.A. Ostroumov have compared the sensitivity of monocellular green algae *Scenedesmus quadricauda* and some higher plants (at early stages of ontogenesis) to detergents. The latter have been found to be more vulnerable to sublethal concentrations of the anionic detergent sodium dodecylsulphate. Moreover, there are many data on differing susceptibility at different stages of ontogenesis.

3.5 Conclusions

Disturbances can manifest themselves at any stage of ontogenesis from embryogenesis to the gerontological processes terminating ontogenesis. In the early stages of ontogenesis — before the birth of higher animals (prenatal stage) —

ontogenetic differentiation may occur at the wrong time (heterochronia), in the wrong place (heterotopia), and abnormal (overdeveloped or underdeveloped) monstrous structure may be formed (teratogenesis). Finally, the whole course of ontogenesis may be disturbed (embryotoxic influence) as a result of the action of anthropogenic factors.

At birth, the disturbances may be concerned primarily with the processes of growth and sexual maturation and propagation. The disturbances given above may be superimposed by a disturbance of the metabolic processes, the development of manifestations of sickness (varying from reversible to lethal), and a change in animal behaviour. As has been shown in experiments, house mice and wild mouse-like hamsters are significantly more susceptible to the influence of chemical pollutants and stress during a period when they are restricted in food and water, that is a situation which occurs frequently in natural conditions for many species. The same study showed the higher susceptibility of natural populations in comparison with laboratory ones (for the same species).

It is possible to make several generalizations.

1. Various physicochemical and other anthropogenic factors acting in the atmosphere, hydrosphere and soil may give rise to disturbance at any stage of ontogenesis and such the disturbances frequently become irreversible.

2. The disturbance of embryonal development and the early stage of post-natal ontogenesis under the influence of practically any anthropogenic factors may occur at significantly lower concentrations (doses) of the acting factor than those resulting in damage to the adult individuals of a given species. There are data on increased susceptibility at the latest (senile) stages of ontogenesis.

3. Threshold concentrations at which the first signs of the action of some factor are observed in the case of chronic action appear to be much lower than under short-range action. It is important to note that over the last decades background concentrations have been rapidly approaching threshold ones (a difference from the 1980's of a factor of 10 or less).

4. Identification of the action of individual pollutants in ontogenesis is complicated due to their complex and continuous interactions, which are sometimes contra-intuitive.

. 5. Environmental pollution may give rise to an overall weakening of species, suppression of the immune system, lowering of resistance to risks from common environmental factors. A reverse phenomenon is also known: sick organisms are more susceptible to the influence of environmental pollution.

6. The injurious action of the majority of pollutants on ontogenesis seems to some extent to be species-specific. There are data on differing susceptibility to embryotoxins or teratogens of genetically different lines within the same species. General rules for this specificity are far from clear.

4 Problems at the Species-Population Level

A population is a basic elementary unit at the species-population level. In the sense of evolution genetics, as population is designated a minimum self-propagating group of single-species individual animals which inhabits a certain space for an evolutionary long period, generates a self-maintained genetic system and forms its own ecological niche. A population is always an adequately numerous group of individual animals which is isolated from other similar groups for a large number of generations.

A basic phenomenon at the species-population level is the directed modification of the population genotype composition, and a basic occurrence is mutation of various types.

Populations and their groups (species) are characterized by their own evolutionary "fate".

For problems of nature protection, the most important fact is that the population level is where evolution takes place, with both disappearance and emergence of the species. Therefore, this is generally the basic level when describing the problems of living nature protection.

Since problems at this level have frequently been reported in the literature, we shall restrict ourselves to a summary presentation and analysis of the subject matter according to three major sections as follows: (1) characterizing species disappearance within large groups of organisms; (2) analyzing the causes of species disappearance; (3) reviewing a number of contemporary species population problems in living nature protection.

4.1 General Characteristics of Species Disappearance

About 1.5 million animal species, 300 to 350 thousand species of higher plants, about 150 thousand fungi and algae species and several thousand prokaryote forms have been described in the scientific literature. According to various estimations, their total accounts for from 1/3 to 1/15 of all species existing on the Earth. Vertebrates are the most studied animals, and for these data exist on rates of extermination by Man over a long period.

Like the emergence of new species, species extinction is a common evolutionary process. However, before Man became a planetary power, biosphere development had followed the path of increasing the "life total" as Charles Darwin called it: extinct species were not simply replaced by new ones, but the number of species had increased during each consecutive epoch of the Earth's development.

The development of mankind has discontinued this natural process. Species that are becoming extinct under the effect of anthropogenic factors are not being replaced by new species, and a significant narrowing of species diversity has come into operation.

The human genus emerged about 3 to 4 million years ago in Africa, and possibly in Southern Asia. The richest fauna in these regions became one of the material bases for the development of the human genus on the Earth, since it allowed Man, a non-specialized predator, the possibility of propagating widely over the entire Earth and eventually creating our human society.

On practically all the continents, but at different times, an epoch of "Pleistocene overhunting" or more exactly "Pleistocene extermination" of the fauna (Table 4.1) was initiated. The only exception is Africa, where the Pleistocene fauna in the form of elephants, rhinoceroses, hippopotami, numerous antelopes and large specialized predators survived better up to now. This paradox may be attributed to the fact that this fauna has evolved concurrently with Man and gradually acquired certain adaptations to coexistence. The rate of the subsequent human propagation on the Earth was too rapid to allow adaptation in animals of other regions.

It is hardly too much to say that hundreds of large animal species had disappeared from the Earth during the Pleistocene extermination epoch. An avalanche-like extinction of small forms was initiated as well (this is much more difficult to prove technically at present) (Owen-Smith 1985).

About 50 mammal species and some 40 bird species have been exterminated by humans in Northern America over the past 3000 years, namely about three

Table 4.1. Total number of genera[a] of large mammals that disappeared in the Pleistocene, apparently exterminated by Man (Romer 1966)

	Africa	Asia	North America	South America	Europe
Bulls, antelopes, deer	23	41	15	5	24
Giraffes, pigs, hippopotami	6	11	2	3	1
Camels	0	1	9	1	1
Horses, tapirs, rhinoceroses	2	6	4	4	7
Hyenas, lions, tigers, bears	4	8	4	2	7
Armadillos, sloths	0	0	8	29	0
Beavers	0	2	4	0	2
Hares, rabbits	1	2	5	0	4
Elephants, mammoths	8	7	8	3	4

[a] The majority of genera consisted of several species.

species per century on the average. In New Zealand, an average of one species has been disappearing every 20 years from the time of the Polynesian settlement of this island in about 950 A.D. to the European's arrival at 1769.

It is possible to recreate the history of species disappearance more exactly since 1600, by using documents, collections and scientific descriptions; therefore, the IUCN has chosen the year 1600 as a reference point for current losses in the world's flora and fauna. About 63 species and 44 subspecies of mammals, 74 species and 87 subspecies of birds have been estimated as having disappeared, which means a loss of more 1.2% vertebrata for this time. These figures are evidently not very precise, but unfortunately underestimated; more vertebrate taxons seem to have died out between that time and the present.

For the present there are no reliable data on rates of decrease of species variability during the last centuries. However, there are some indicative data for Northern America, where the levelling-off rate of species disappearance occurred between the Pleistocene extermination and European emigrant settlement.

Subsequently, the rate of species extinction has accelerated (Fig. 4.1). One or two species seems to have died off in the 17th to 18th centuries. This rate has risen to several species of vertebrates per year in the 20th century (an average of one species had been dying every 3 years for 400 years). At present the extinction rate is rising. According to certain estimates (Myers 1988) recently about 4 to 27 species (including vertebrates and invertebrates and plants) are disappearing every day. By the end of the 1990's, one species or more can disappear every hour.

Approximately 2.5 to 3.0% of higher vertebrate species and about 10% of higher plant species are now in danger of disappearance.

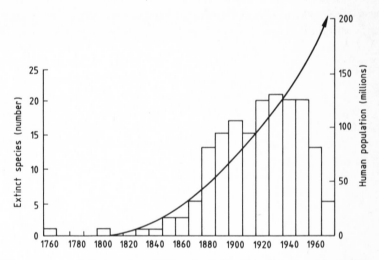

Fig. 4.1. Rate of disappearance of plant species in US flora (excluding Hawaii) by decades over 200 years. The sharp decrease in rates over recent decades is an artefact because one is only sure of the complete disappearance of a species many years after the last observation. The *solid line* shows the population increase in the United States over the same period

According to the global forecasts, which are based on the rate of decreasing natural habitats within the most species-rich regions of the world (in the tropics), from 500,000 to 1 million species (mainly plants, fungi, mushrooms and other small organisms belonging to various taxons) may be in danger of extinction by 2000. Some forecasts suggest the possible disappearance of 10 to 20% of all existing species by 2000.

Let us consider the materials available for individual groups in more detail.

4.1.1 Problems in the Protection of Prokaryotes, Fungi, Mushrooms and Lichens

These problems are extremely important in view of the rapid progress in biotechnology.

4.1.1.1 Problems in the Protection of Prokaryotes

The total number of prokaryote species described, including cyanobacteria, is several thousands. Their actual number in the biosphere is considerably higher; according to the view of G.A. Zavarzin (personal communication), a very high number of prokaryotes not described elsewhere inhabit the invertebrates as symbionts. A very significant number of species and strains of prokaryotes, otherwise unknown, inhabit diverse soils, sludges and natural waters, as well as being associated with plant or animal organisms, this association being frequently species-specific and obligatory. Therefore, the problem of protecting prokaryote variability lies mainly in the maintenance of their habitats, i.e. areas of different soil types with indigenous ecosystems, including complete sets of species of plants, mushrooms, animals, and natural aquatic reservoirs with all hydrobionts and undisturbed shores and bottom.

Present work does not solve the acuteness of the problem, since, firstly, it covers only the known prokaryote species and a very great number of additional species inhabit natural habitats, and, secondly, the scientific collections of specimens of prokaryotes do not cover the total variability of natural strains among species which were described and are represented in these collections. However, specific strains are very important for natural environmental processes (e.g. degradation of certain stable natural compounds such as lignine) and for biotechnological purposes.

4.1.1.2 Fungi

About 100,000 true fungi and some 500 species of mixomycetes are known. Exact data on the number of endangered fungi species are insufficient. For example, it is known that in some parts of Central Europe, out of 3000 true fungi species over 1000 species are assignated as rare and in need of special protection. The disappearance or at least the abrupt decrease in the number of the majority of edible fungi species in Europe is well known.

Among 2337 macromycete species of the groups Boletales, Agaricales, Russulales and Gasteromycetidae in the FRG about 23 species (0.1%) are extinct or have disappeared, about 345 species (14.8%) are in acute or great danger, another 345 species (14.8%) are less endangered, another 147 species (6.3%) are potential endangered, i.e. a total of 860 species (36.8%) are in danger of varying severity. In general in macromycetes (Großpilze), about 37 species are now extinct or have disappeared in the FRG and a total of 1037 species are in danger (including potentially endangered species) (Blab et al. 1984).

In Sweden at least 50 fungi species are estimated to be in danger. Two species of Ascomycetes and 18 species of Basidiomycetes are among the species in the USSR Red Data Book (RDB USSR, 1984b).

For protection of the genetic pool of fungi at least three groups of problems are significant: (a) the vagueness of the concept "geographic range" and "endemism" for an enormous number of mushrooms; (b) specialized symbionts and parasites; (c) mycorrhiza-forming mushrooms with large fruiting bodies.

The vagueness of the concept "geographic range" for a great multiplicity of small soil fungi, which are frequently named "micromycetes", is due to the fact that identifying these fungi is open to ambiguous interpretation. As a result, for a significant number of fungi there are no data on the limits of geographic ranges and on the extent of endemism of a particular species.

Many fungi species are highly specialized symbionts or parasites. They are also frequently obligatory symbionts or parasites and have no ability to develop normally in the absence of their respective partner, which can be a plant, another mushroom, arthropod or another organism. Such a partner is a requisite component of the ecological niche. Thus success in the protection of a significant part of the fungi species is directly dependent on the maintenance of populations of other organisms which inhabit concurrently with them and form the total ecological cluster.

Problems in the protection of mycorrhiza-forming mushrooms with large fruiting bodies (all edible mushrooms are included in this category) are similar to the problems in the protection of highly decorative flowering plants. Excessive harvesting of mushrooms in suburban areas results in the extermination of many populations. Anthropogenic variation of forests, substituting secondary forests for primary ones (e.g. leaf forests for coniferous ones) results in a sharp decrease of geographic ranges for many mycorrhizal mushrooms and eventually in the disappearance and displacement of these mushrooms by phytocenoses.

4.1.1.3 Lichens

About 26,000 species of lichens are known. These organisms, having a taxonomic level of division or type, play an important role in ecosystems and participate in the soil formation process and in building up moss and lichen synusias in forests, mountains and tundras. They are producers of many substances which are biotechnologically important.

The problems in lichen protection are similar in many respects to those for fungi or mushrooms. Among them there are numerous rare or endemic forms, species of decreasing geographic ranges, and with diminishing populations due to pollution (especially of air) or too intense harvesting.

About 29 of the lichen species in the USSR are included in the USSR Red Data Book (1984b).

In the FRG, out of 1850 lichen species, some 26 species are extinct or have disappeared (more than 1%), about 246 species (approximately 15%) are threatened with extinction, and an additional 108 species are endangered (but to a lesser extent), 35 species are in "potential danger". A total of 35% of lichen species are considered to be endangered to some extent (Blab et al. 1984).

At least 64 lichen species are endangered in Sweden (Ingelög 1985).

4.1.2 Plants

We consider questions concerning (1) Algae, (2) Bryophyta, and (3) vascular plants.

4.1.2.1 Algae

The problems in maintaining the variability of algae and revealing the extent of the threat to their genetic pool have received hardly any attention. Until now it is unknown how many of the total number of algae species such as Chlorophyta (total 6000 species), Phodophyta (3000 species), Chrysophyta, Diatomeae and Xanthophyta (in all 10,000), Phaeophyta (about 1000 species), Pyrrophyta (more than 1000 species), Euglenophyta (more than 300 species) are under threatened, which is an alarming fact in itself, especially considering the value of the algae genofond for biotechnology.

The fact that many algae species are rare can increase their vulnerability and suggests that protecting them is expedient.

Of 900 seaweed species to be found on the shorelines of Great Britain, about one third is considered to be rare.

The overwhelming proportion, about 82.4% stonewort species (Charophyta) in the FRG (34 species of Charophyta have been found in the FRG), are threatened to some extent. Two species (i.e. 5.9%) were extinct or have disappeared, a further two species are on the verge of extinction and 24 species are threatened. Only six stonewort species are considered to be beyond danger in the FRG. All the freshwater species of brown algae (four species) and red algae (28 species) in the FRG are included in the so-called Rote Liste (Blab et al. 1984), the equivalent to the RDB, which means that 100% of the species of these groups is threatened to some extent.

4.1.2.2 Bryophyta

About 87 Bryophyta species are included in Sweden in the list of rare or disappearing species. Many species are to be found at only three to ten locations in Sweden, in spite of the apparent abundance of their habitats. In Great Britain the mosses *Helodium lanatum* and *Paludella squarrosa* (both bog mosses) have disappeared for the past decades. Air pollution has resulted in the disappearance of epiphitic bryophytes of *Orthotrchum* genus in the greatest part of Britain. Excessive harvesting has resulted in the disappearance of the moss *Cyclodictyon laetevirens* from its sole location in England (Mousehole Cave in Cornwall) as far back as some 50 years ago. A significant amount of tropic Bryophyta are forest species and therefore their survival depends on maintaining the tropic forests.

Out of about 1000 Bryophyta species in the FRG, 15 are extinct or have disappeared, 84 species are under threat of disappearance, a further 40 species are in the "potentially threatened" state. Among approximately 550 species which have been found in Schleswig-Holstein, about 16% have disappeared already and about 45% are seriously threatened; only 25% of the species are out of danger of a sharp decrease in number (Blab et al. 1984).

Thirty two Bryophyta species are included in the USSR Red Data Book (1984).

4.1.2.3 Vascular Plants

At least 25,000 to 30,000 vascular plant species, i.e. about 8 to 10% of the total number of all vascular plant species described on the planet, are under threat of extermination or can become critical in the near future.

In certain regions several percent of the flora have disappeared over the past 20 years (Table 4.2).

The extinction rate is increasing (e.g. Davis et al. 1986). Thus, in the USA the number of species which became extinct in the second half of the 19th century have increased tenfold in comparison with the first half of the same century.

For a number of species which are represented by one single specimen the situation is catastrophic (Table 4.3).

Table 4.2. The scale of extinction of vascular plants during the 19th and 20th centuries. (Malyshev 1981; Yablokov and Ostroumov 1985; Blab et al. 1984 and others)

Region	Rate of extinction, number of species	% from flora
USA, continental	ca. 100	ca. 0.5
USA, Hawaii	255–270	ca. 11.6
Holland	50–75	ca. 4.0
Belgium	62	4.8
Sweden (from 1850)	37	2.3
GDR	29	1.6
ČSSR	76	ca. 4.0
France (from 1930)	40	ca. 1.0
FRG	60 (vascular)	2.4
	15 (Bryophyta)	1.5–2.0
Hungary	23	1.1
Bulgaria	31	0.9
Switzerland	46	?
Australia	76	?

Table 4.3. Number of known vascular plant species maintained in nature in a small number of specimens [minimal estimate according to the IUCN Red Data Book (1978) with additions]

Number of specimens remaining in nature	1	2	3	4–5	6–21
Number of species with given number of specimens	10	1	2	4	6

Table 4.4. The number of endangered and threatening species of vascular plants in some regions. (Malyshev 1981; Yablokov and Ostroumov 1983, 1985; Blab et al 1984; Leigh et al. 1984; Cerovsky 1985; Landol et al. 1982; Ingelög 1985 and others)

Region	% Endangered and threatened species	Number of endangered and threatening species	
Tropical zone	ca. 30		
Temperate zone	ca. 5		
USA, Hawaii	43		
continental	ca. 10		
FRG (Spermatophyta +			
(Pteridophyta)	35	697	(1982)
(Bryophyta)	14	99	(1984)
Switzerland	ca. 28	940	(1982)
USSR	ca. 20		
Siberia	4.4		
Karelia	16.4	180	(1979)
Moldavia	13.7	240	(1980)
Pskov district	12.3	160	(1980)
Riazan district	5.1	58	(1981)
Bulgaria	ca. 21	732	(1984)
Australia	Minimum 10	2206	(1981)
Hungary	More than 14		
Sweden	More than 5	73	(1985)
GDR	More than 7	133	(1984)
Greece	ca. 20	750	(1979)
World total	ca. 17	ca.42,0000	(1980)

The situation for many flora according to the number of the threatened species is alarming (Table 4.4).

About 603 vascular plant species are recorded in the USSR Red Data Book (1984b). Representatives of the following families are the most numerous (total number of species included in the RDB and the number of the category 0 species are in parentheses): *Allium* (12), *Amaryllis* (11), Umbrelliferae (39; 2); Compositae (53; 2); Brassicaceae (20; 1); Campanulaceae (17); Caryophyllaceae (17; 1); Ericaceae (12); Leguminosae (36; 4); Iridaceae (19; 1); Labiatae (14; 5); Liliaceae (43; 3; 11 species of *Tulipa*); Orchidaceae (35; 1); Paeonia (8); Gramineae (23; including four species of wheat and two species of rye);

Primulaceae (16; 1); Rosaceae (17; 2; the 17 species including two species of almond, three species of pear; one species each of cherry, strawberry, apple, and plum).

4.1.3 Disappearing Invertebrates

More than 1.5 million invertebrate species have been described. The IUCN Red Data Book on invertebrates (1983) includes several hundreds of taxons and far from all species of endangered invertebrates have been considered. Subsequent volumes of this RDB are intended to cover data for disappearing Lepidoptera species (the volume on the swallowtail family (issued in 1984 includes 155 species of these butterflies) and mollusks, as well as for communities in caves and coral reefs. The actual number of disappearing invertebrates appears to be measured by tens of thousands. For example, on the Hawaii Islands, out of 1061 endemic mollusk species about 600 species appear to be already extinct and up to 400 species are under threat of disappearance. The disappearance rate for insects or other invertebrates has been shown to be especially high in the industrially developed regions of the world. Thus, in North America some 1000 mollusk species are described, of which about 40 to 50% are now either extinct or are under threat of disappearance. Two thirds of the butterfly species of Europe at present are under threat of disappearance. Four butterfly species disappeared in the Netherlands between 1946 and 1980. In Great Britain, four dragonfly species (about 10% of all species) have become extinct over the last 10 years. In the territory of the FRG, 27 species of large butterflies have been extinct for the past years. An appreciable decrease in the number of butterfly species is evident in some regions of the USSR also. For example, out of 150 species of daylight butterflies in Western Alatau, 8% had disappeared and 18% had become very rare.

The analysis of the proportion of endangered species in various taxons (Table 4.5), conducted in the FRG, has shown very alarming picture. About one half of the genofond of echinoderms, freshwater mollusks, Neuroptera, caddis flies, dragon flies, may flies, Decapoda, crayfish, more than 1600 Coleoptera species, more than 600 Hymenoptera species, more than 490 species of large butterflies, all the genofold of Echiuriodea, as well as the Anostraca, Notostraca and Conchostraca are endangered.

4.1.4 Disappearing Fishes

According to the IUCN Red Data Book (1978) 168 species (0.84% of the world's fauna) and 25 subspecies of fishes are under threat of extermination. For the freshwater fishes, the number of endangered species approaches 3.5%. Among the freshwater fishes of Europe, 52.3% of the species are under threat of disappearance in certain parts of Europe (Lelek 1980) and at least one species (*Coregonus oxyrhynchus*) is threatened by complete extermination (only about 2000

Table 4.5. Extent of threatening the genetic pool of some taxons of invertebrata in FRG. (According to Blab et al. 1984)

Taxon	Number of species			
	Total species in fauna of the FRG	Disappeared or not found for a long time	Under threat (including disappeared species)	Under potential threat
Echinodermata	37	4 (11%)	4 (11%)	15 (41%)
Bivalvia	31 (freshwater)	1 (3%)	10 (32%)	7 (23%)
Gastropoda	270 (freshwater)	2 (1%)	58 (21%)	70 (26%)
Hemiptera	800	11 (1%)	41 (5%)	—
Thysanoptera	222	2	9 (4%)	16 (7%)
Some taxons of Hymenoptera (Symphyta, Scolioidea, Formicoidea, Pompiloidea, Vespoidea, Sphecoidea, Apoidea)	1686 (in taxons analyzed)	58 (3%)	615 (36%)	—
Macrolepidoptera	1300	27–28 (2%)	494–507 (38%)	40 (3%)
Trichoptera	278	19 (7%)	122 (44%)	46 (17%)
Some taxons of Diptera (Psychodidae, Ceratopogonidae)	About 600 (in families analyzed)	48 (8%)	176 (29%)	23 (4%)
Mecoptera	8	1 (13%)	1 (13%)	—
Neuroptera	103	—	45 (44%)	7 (7%)
Some taxons of Coleoptera beetles (total number of beetle species in the FRG is 5727 species)	4073 (in taxons analyzed	96 (2%)	1610 (40%)	76 (2%)
Orthoptera s.lat. (Blattodea, Mantodea, Saltatoptera, Dermatoptera	97	5 (5%)	36 (37%)	—
Plecoptera	119	12 (10%)	44 (37%)	—
Odonata	80	4 (5%)	43 (54%)	—
Ephemeroptera	81	5 (6%)	49 (60%)	8 (10%)
Decapoda	63	1 (2%)	4 (6%)	28 (44%)
Some taxons of Phyllopoda (Anestraca, Notostraca, Conchostraca)	10 (in taxons analyzed	3 (30%)	10 (100%)	—
Araneae	803	17 (2%)	100 (12%)	14 (2%)
Opiliones	39	—	4 (10%)	1 (3%)
Echiurioidea	1	—	1 (100%)	—

specimens were found in 1984). In the seas surrounding Europe, 53.5% of the populations of all commercial fishes are assigned to the category of depleted or completely prohibited for any trade that is associated with an abrupt decrease in number. About 33.3% of all fish species of the Cape Province in South Africa and 26.9% of fish species in Czechoslovakia are considered to be disappearing. In the FRG, of 70 species of freshwater fishes and Cyclostomata, four species had disappeared or are not identified; a total of 50 species (71%) are endangered (Blab et al. 1984).

The USSR Red Data Book (1978) includes nine fish species (3.5% of overall number of freshwater forms), but in individual regions the situation is more dangerous. For example, in Tadzhikistan, 10.2% of all fish species and in the Ukraine 20.0% of all fish species are endangered. In the Gorki District 36.8% of local species (21 species out of a total of 58) have disappeared over the past decades. In many locations the number of previously high-volume species have diminished by several times. Thus, the number of inconnu (*Stenodus nelma*) and whitefish *Coregenus muksun* have decreased considerably. The ichtiofauna of the Aral Basin is in the greatest danger, since hydrotechnical engineering work in Central Asia is calling the very existence of this basin into question. Additional data see in (Allendorf 1988).

4.1.5 Disappearing Amphibians and Reptiles

Two amphibian species became extinct in historic times by human action, and one species (the Israel spotted frog, *Discoglossus nigriventer*) has been extinct since the 1950's. According to the obviously underestimated data of the IUCN Red Data Book (1979), at least 33 amphibian species (about 2% of the world's fauna) are endangered. In the FRG 58% of amphibian species (11 of 19 species) are threatened (Blab et al. 1984). Thus, in Westfalia, for example, there remain less than 100 populations of amphibian species common until recently, such as *Bombina variegata, Hyla arborea, Pelobates fuscus, Rana ridibunda* and *Rana arvalis.*

In the USSR Red Data Book (1984a) are included 60% species of tailed amphibians and 13% of *Anura* species, i.e. 27.3% (9 of 33) of the total species in this group of vertebrata.

About 20 reptile species have become extinct in the world since 1600. On the present average, at least 1.8% of all the world's reptile species are endangered. However, for certain regions the situation is more alarming. In Austria and Czechoslovakia about one fourth of the reptile species is endangered. In the FRG, about 75% of species (9 out of 12 have been placed on the Rote Liste (Blab et al. 1984).

Two families of reptiles are under threat of disappearance: leatherback turtles (Dermochelidae) and gavials (Gavialidae) and two further families (sea turtles Chelonoidae, from six species) and true crocodiles (Crocodylidae) of 13

Table 4.6. Distribution of rare and disappearing amphibians and reptiles of the USSR fauna. (USSR RDB and the IUCN RDB 1979)

Order	Total number of species in the USSR	In the USSR RDB		Number of threatened species of world fauna (RDB IUCN 1979)
		1978 Ed.	1984 Ed. (RDB USSR 1984a)	
	Reptiles			
Tortoises	7	2 (28.6)[a]	2 (28.6)	26 (13.0)
Lizards	77	10 (13.0)	19 (24.7)	31 (0.9)
Serpents	60	9 (15.0)	16 (26.7)	25 (1.3)
All reptiles	144	21 (14.6)	37 (26.7)	105 (1.8)
	Amphibians			
Caudata	11	5 (45.5)	6 (54.6)	19 (6.8)
Ecaudata	23	3 (13.0)	3 (13.0)	22 (1.2)
All amphibians	34	8 (23.5)	9 (26.5)	41 (2.0)

[a] Percent of total number of species are in parenthesis.

species are under threat. About eight genera of reptiles are among the disappearing animals.

In the USSR two species (28.6%) of tortoises, 19 species (26.4%) of lizards and 16 species (26.7%) of serpents are in the threatened state; a total number of 25.7% of reptile species has decreased dangerously (Table 4.6).

4.1.6 Disappearing Birds

About 109 bird species (approximately 1%) have become extinct on the Earth since 1600 and at present more than 3.0% (274 species) are endangered; according to other data, 503 bird species were endangered in 1982.

One suborder (kagou, Rhinocheti), one family (shrubbery birds Atrichornithidae) and 17 various genera are under threat. Within other groups cranes, bustards, all predator birds, pheasants, parrots and birds of paradise have become most endangered (according to the number of endangered species).

Until 1981 there were at least four species in the world with a number of specimens of from two to five, five species with a number of specimens up to 20, 12 species from 25 to 50, 5 species from 55 to 100, 10 species from 108 to 300 (Table 4.7). Taking into account the difficulties in precise recording of birds, it is evident that several bird species now have less than 250 specimens, and some 20 species less than 50 specimens. Even with this low general level, the number of specimens is decreasing rapidly. Thus, the number of California condors is decreasing from initially 25 specimens (1980) by three to four specimens per year; the number of Mauritius parrots diminished from 20 to 5 specimens after a severe hurricane in 1979.

Out of 408 bird species of Western Europe, 294 species are endangered (72%). About 8% of species nesting until the 1930's in FRG territory (20 species out of

Table 4.7. The rarest birds. (International Red Data Book 1982, King 1981 and others)

Species, subspecies	Location	Number and year
Madagascar serpent eagle, *Eutriorchis astur*	East Madagascar	Last observation early 1950
Imperial woodpecker, *Campephilus imperialis*	Mexico	Last observation 1958
Marianas mallard, *Anas oustaleti*	Guam Is., Hawaii in captivity	3 (1975), 2 (1982)
Kanai O'o' *Moho braccatus*	Hawaii in captivity	2 (1981)
Reunion petrel *Pteroderma aterrima*	Reunion Is., Rodriges Is.	Several examples (1974)
Mauritius parakeet *Psittacula crameri echo*	Mauritius Is.	24 (1976) 5 (1981)
Laurel pigeon, *Columba junoniae*	Canarias Isl.	8 (1977)
Japanese crested ibis, *Nipponia nippon*	China, Japan in captivity	13 (1983)
Mauritius kestrel, *Falco punctatus*	Mauritius Is.	15 (1982)
Madagascar sea eagle *Haliaeetus principalis*	Madagascar	ca. 20 (1981)
Ivory-billed woodpecker, *Campephilus principalis*	Cuba, Florida	Less than 20 (1981)
California condor, *Gymnogyps californianus*	USA, in captivity	15 (1987)
Aldabra brush warbler, *Nesillas aldabrans*	Aldabra Is.	ca. 25 (1981)
Lord Howe wood rall, *Tricholimnas sylvestris*	Lord Howe Is.	ca. 30 (1981)
New Zealand black stilt, *Himantopus himantopus*	New Zealand	ca. 30 (1981)
Kakapo, *Strigops habroptilus*	New Zealand	ca. 40 (1981)
Seyshelles magpie robin, *Copsychus seishellarum*	Fregate Is.	ca. 40 (1981)
Rodriges brush warbler, *Bebronis rodericana*	Rodriges Is.	Less than 50 (1981)
Okinawa woodpecker, *Sapheopipo noguchii*	Okinawa Is.	ca. 50 (1981)
White-breasted silvereye, *Zosterops albogularis*	Norfolk Is.	Less than 50 (1981)
Mauritius pink pigeon, *Nesoenas mayeri*	Mauritius Is.	ca. 50 (1981)
Chatham island oyster-catcher, *Haemotopus chatamensis*	Chatam Is.	ca. 50 (1981)
Seychelles black paradise flycatcher, *Terpsiphone corvina*	Seychell Isl.	60–70 (1981)
Noisy scrub bird, *Atrichornis clamosus*	Australia, (one forest)	80–90 (1981)
Atitlan grebe *Pochlimbas gigas*	Guatemala (one lake)	ca. 100 (1981)
Kabyllan nuthatch, *Sitta ledanti*	Algiers (one forest)	108 (1981)

Table 4.8. Distribution of rare of disappearing bird species of the USSR, listed according to order. (RDB of the USSR and IUCN)

Order	Total number of species in the USSR	In the USSR RDB		Number of endangered species of the world's fauna
		1978 Ed.	1984 Ed.	
Procellarii-formes	18	1 (5.6)[a]	3 (16.7)	13 (16.0)
Pelecaniformes	11	2 (18.2)	3 (27.3)	4 (8.0)
Ciconiiformes	23	9 (39.1)	6 (26.1)	11 (9.8)
Anseriformes	57	11 (19.3)	11 (19.3)	13 (9.2)
Falconiformes	43	14 (32.6)	18 (41.9)	24 (8.9)
Galliformes	20	6 (30.0)	6 (30.0)	37 (14.2)
Gruiformes	23	8 (34.8)	10 (43.5)	25 (14.2)
Charadriiformes	133	13 (9.8)	14 (10.5)	14 (4.7)
Columbiformes	17	2 (11.8)	2 (11.8)	22 (7.5)
Strigiformes	17	1 (5.9)	1(5.9)	12 (9.0)
Piciformes	14	1 (7.1)	1 (7.1)	11 (2.9)
Passeriformes	312	3 (1.0)	5 (1.6)	135 (2.6)
Total birds	765	63 (8.2)	80 (10.4)	396 (4.6)

[a] Percent of total number of species is in parentheses.

255) had disappeared by 1984 and the share of the endangered species is of 31% (78 species); in addition, other 35 species are potentially endangered (Blab et al. 1984).

About 80 bird species, i.e. 10.4% of the total aviafauna of the USSR, are included in the RDB (Table 4.8). Within certain regions tens of species of nesting birds have disappeared over the last 50 to 100 years. At least three bird species (the Japanese crested ibis, *Tadorna cristata*, and the woodpecker, *Picus squamatus*) appears to have disappeared from USSR fauna over the last 30 years and for at least six species, no more than 100 specimens have been recorded since 1980.

4.1.7 Disappearing Mammals

Some 64 mammal species have been exterminated in the world between 1600 and 1974, and 233 species (6.2%) are now endangered. There are the endangered orders Sirenia and Proboscidea (all the species are included in the IUCN Red Data Book), as well as the families Daubentoniidae and Solenodontidae and the anthropoid apes. There are about 30 disappearing mammal genera in the world. Eight mammal forms are now found only in captivity and do not exist in nature (see Chap. 11), and the number of at least 20 further species does not exceed 200 specimens. Among them there are the kouprey (*Bos sauveli*, Thailand, Kampuchea), deer *Cervus albirostris* (Tibet), tamarao (*Bubalus mindorensis*, Mindao Island, the Philippines), Javan rhinoceros (*Rhinoceros sondaicus*, Indonesia), pygmy swine (*Sus salvelinus*, India, Bangladesh), monkeys *Leontopithecus chrysomellas*, *L. chtrysopygus* (Brazil), red wolf (*Canis rufus*, USA), cat *Felis iriomotensis* (Japan), Vancouver marmot (*Marmota vancuverensis*, Canada) and others.

More than half the mammals of Europe (155 species) are endangered (55 species are especially endangered). In certain parts of Europe more than half the mammal species have already disappeared (Feldman 1984). In the FRG 47% of the species (44 species) are in the RDB (Blab et al. 1984).

In USSR territory, at least six subspecies, seven species, two genera, two families and one order of mammals (Table 4.9) disappeared from the fauna between the 16th and the mid-20th century. At least four animals (monk seal, cheetah, Turan tiger, Mongolian gazelle, and possible the red wolf) have disappeared from the USSR fauna over the past 15 years.

At present 21.8% of the mammal species have become endangered in the USSR. Ten mammal species have no more than 200 specimens (Frontal Asian leopard, Eastern Siberian leopard, striped hyena, polar fox of Mednyi Island, Kyzyl Kum wild sheep, Greek mole rat, Ussuri tiger, Altai beaver, Korean grey whale, Southern Okhotsk right whale).

4.2 Causes of Species Disappearance

The data given in the last section of this chapter suggest that at least 2% of the present amphibian and reptile fauna, 3.5% of freshwater fishes, almost 5% of birds, more than 6% of mammals and about 10% of vascular plant species of the world are endangered.

What are the causes for this situation? Analysis of the causes for the disappearance of 194 vertebrate species from the Earth between 1600 and the present has shown that on an average for all the groups, the principal causes of extinction are the influence of introduced species, immediate harvesting (hunting) and destruction of habitats (Tables 4.10, 4.11).

These data are certainly not precise, since one and the same species is generally influenced by several factors diminishing its number. In the calculations shown, the influence of the principal factor effecting the decrease of certain species was assumed to be the only one.

The situation is, of course, dynamic and can change rapidly. Thus, according to data for the FRG, direct victimization (hunting, collecting, nest destruction, etc.) in 1972 directly influenced more than half of the endangered species, but by 1980 only 28% of the endangered species decreased in number due to these causes and the negative effect of intensifying agricultural production and occasional loss (e.g. due to vehicles) sharply increased.

To present the scales and the nature of action of these causes more clearly, let us consider the influence of selected factors more comprehensively.

Table 4.9. Species and subspecies of USSR mammals disappeared over the last 1500 year. (Smirnov 1983; Yablokov 1985 and others)

Species, subspecies	Time of last observation in the USSR	Level of the loss for	
		USSR fauna	world fauna
Wild camel, *Camelus bactrianus*	4th century	Order	—
Tibetian antelope, *Pantholops hodsoni*	8th century	Species	—
Transbaikalis bubalus, *Parabubalis capricornis*	10th century	Species	Species
South Siberian bison, *Bison priscus* spp.	10th century	Species	Species
European forest ox, *Bos primigenius longicornis*	16th century	Subspecies	Subspecies
Baltic grey whale, *Eshrichtius robustus* spp.	16th century	Subspecies	Subspecies
Japanese-sachaliensis grey whale, *Eschrichtius robustus* spp.	17th century	Subspecies	Subspecies
Sibirian steppe bison *Bison priscus diminutus*	18th century	Species	Species
Forest wild horse, *Equus cabbalus silvaticus*	18th century	Subspecies	Subspecies
Baikal yak, *Bos (poephagus) baikalensis*	18th century	Species	Species
Steller sea cow, *Hydrodamalis stelleri*	1768	Order	Family
Kazach wild ass, *Equus hemionus finschi*	19th century	Subspecies	Subspecies
Steppe wild horse, *Equus caballus gmelini*	1918	Species	Species
Mongolian wild ass, *Equus hemionus hemionus*	1926	Subspecies	Subspecies
Black sea monk seal, *Monachus monachus*	1968	Family	—
Turan tiger, *Panthera tigris virgata*	1971	Subspecies	Subspecies
Mongolian gazelle *Procapra gutturosa*	1970	Genus	—
Cheetah, *Acinonyx jabatus* ssp.	1979	Genus	Subspecies

Table 4.10. Number of vertebrate species (%) which became extinct due to various causes between 1600 and 1974. (Overall data)

Cause	Amphibians	Reptiles	Birds	Mammals	All groups
Gathering, harvesting or hunting	0	32 (7)[a]	19 (21)	23 (14)	21.5 (42)
Destruction of habitats	100 (1)	5 (1)	20 (22)	19 (12)	18.3 (36)
Effect of introduced species	0	42 (8)	22 (24)	20 (13)	23.0 (45)
Direct extermination	0	0	0	1,6 (1)	0.5 (1)
Occasional loss, disease etc.	0	0	1	0	0.5 (1)
Natural factors	0	0	1	1	1.0 (2)
Unknown	0	21 (4)	37 (40)	36 (23)	34.3 (67)

[a] Absolute number of species is in parentheses.

Table 4.11. Number of endangered vertebrate species (%). (Nilsson 1983)

Cause	Fishes	Amphibians	Reptiles	Birds	Mammals	All groups
Excessive procurement (hunting, collecting)	9.8(16)[a]	5(2)	50(39)	20(54)	31(72)	23.2(183)
Destruction of habitats (beyond pollution	65.9 (111)	82(27)	22(16)	59(161)	32(74)	49.5 (389)
Effect of introduced species	23.8 (40)	9(3)	24(19)	12(33)	17(40)	17.1 (135)
Direct extermination	0	0	2(2)	0.6(2)	8(19)	3.0(23)
Occasional loss, disease, etc.	0.5(1)	0	1(1)	1(3)	2(5)	1.3(10)
Natural factors	0	0	0	1(3)	0	0.4(3)
Pollution	0	4(1)	0	1(3)	0	0.5(4)
Disturbance	0	0	0	2(6)	0	0.8(6)
Unknown	0	0	1(1)	3(9)	10(23)	4.2(33)

[a] Absolute number of species is in parentheses.

4.2.1 Destruction of Habitats

This has become the cause of decrease in number and extinction of many insect species. Thus, pasturing livestock endangers the wingless locust (*Achurimima* sp., Australia) which is known only in six small populations, as well as the Australian grasshopper, *Keyacris scurra*. The latter species have disappeared from many biotopes when the intense pasturing of sheep resulted in the eradication of phytocenoses of the kangaroo grass, *Themeda australis*. In the USA, pasturing sheep endanger the populations of the butterfly, *Boloria acrocnema*.

Direct extermination of habitats due to urbanization has resulted in the extinction of three endemic sand dune butterflies, *Cercycnis sthenele sthenele*, *Glaucopsyche xerces* and *Icaria icaroides pheres*, which have disappeared as a result of the growth of San Francisco and Los Angeles.

Most of the species that have disappeared as a result of habitat destruction inhabit ecosystems associated with freshwater, tropic forests and islands, mostly in North, Central, and South America, Southeastern Asia, Madagascar, the Carribean Islands, the western part of the Indian Ocean and Africa. Of 15 most endangered species of European butterflies eight species have wetland habitats.

Six extensive species of singing Passeriformes in the USA have decreased by 1 to 4% annually from 1968 until 1980. This decrease can be correlated with the diminishing of forest areas in the locations where these birds winter in Central and South America.

The destruction of habitats appears to be one of principal causes of the dangerous decrease in the number of species endemic to the USSR such as the Russian muskrat. This insect-eating animal with an excellent fur inhabits the shores of quiet, clear rivers in the central area of the European part of the USSR. The commercial development of lands leaves increasingly few of such locations and the number of muskrats is dropping catastrophically every year.

Usually the first sign of the effect of habitat destruction is insularization, i.e. the disintegration of a previously single geographic range to small islands (Fig. 4.2).

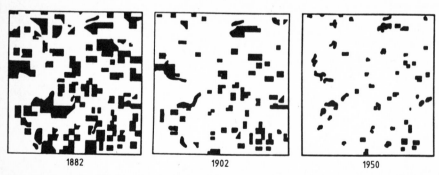

1882   1902   1950

Fig. 4.2. Example of habitat insularization. All this territory (625 ha) in Wisconsin (USA) was completely covered with forest. The *black spots* are forest-covered areas. (After Pianka 1978)

Species disappearance as a result of disturbing the habitat occurs the more quickly, the less area of inhabited islands remains and the more they are isolated from each other. Thus, methods of theoretical biogeography have calculated an expected rate of decrease in the number of large animal species in 19 national parks of East Africa such as Serengeti, Cabalega Falls, etc. It has been found that insularization could result in the loss of 11% of the species of large mammals after 50 years and of 44% after 500 years (Soule 1980).

At least one third of the rare or disappearing plant species in the USSR are suffering from land being used as pasture, and other forms of habitat destruction.

The following gives the rate of destruction of plant habitats which are worth protection. In 1975 in the FRG a list of about 50,000 biotopes was compiled which should be protected. In 1982 out of them twenty thousand biotopes have been destroyed or changed to such an extent that were not able to maintain the existence of species valuable enough for protection. In the FRG, the principal threat for 63% of all disappearing plant species is the destruction of habitats resulting from development of agriculture. The plant that has the largest flowers in the world (*Rafflesia arnoldi R.br.*), parasite on the tropical liana of the genus *Tetrastigma*, is also on the verge of disappearance due to the destruction of its habitat, i.e. the tropical forests in Sumatra (Indonesia).

Habitat destruction is one of the most important factors endangering rare or disappearing species of the USSR flora, along with excessive procurement or harvesting (Table 4.12). It is emphasized that in many cases habitat destruction acts concurrently with other factors.

Table 4.12. Principal families of higher plants of the USSR for disappearing species of which the destruction of habitats or excessive harvesting is or was a leading factor in decrease of number. (Calculated according to data of various authors)

Family	Number of rare or disappearing species or subspecies		
	Total taxons in the USSR	% Suffering from the destruction of habitats	% Suffering or having suffered from excessive harvesting
Amaryllidaceae	5	20	80
Araliaceae	5	20	80
Cupressaceae	6	50	83
Ericaceae	12	33	58
Leguminosae	23	52	8
Fagus	6	33	67
Iridaceae	18	28	67
Liliaceae	22	32	77
Orchidaceae	35	23	75
Pinaceae	9–12	ca. 17	100
Polygonaceae	7	14	57
Primulaceae	15	13	53
Thymelaeaceae	6	100	0

4.2.2 Over-exploitation

By exploitation is meant both hunting and any other withdrawal or procurement of animals or plants from the environment to collect, maintain in captivity, make souvenirs, produce wood stock, seeds as raw materials for medicinal use, for use in tests, etc. Such over-exploitation is the second important cause of the dramatic decrease in animal (and some plant) number.

Among species which have decreased in number due to this cause are sturgeon fishes and other commercial fishes in certain regions, sea turtles, land turtles, large Ungulata, crocodiles, many monkeys, some game birds (ducks, pheasants), otters, cats, parrots, as well as cactuses and other succulents, some drug plants, food plants, ornamental plants, etc.

4.2.2.1 Mammals

The scale of killing game animals and trade mammals is very significant. However, the trade has not always resulted in diminishing numbers. Thus, the European countries (including the European part of the USSR) were inhabited by about 8.41 million Ungulata in the mid-1970's, of which 2.32 million specimens were shot every year. In this situation both the number, and the procurement volume for most Ungulata species have permanently increased. In European countries, about 6.84 million specimens of red fox were killed between 1971 and 1978 (more than 860 thousand specimens per year). The total number of moles killed annually in the 1970's in France and the USSR alone was about 7.39 million specimens; more than 7.1 million specimens of other fur animals in Europe were killed annually (Dezhkin 1983). In Australia, about 25 million kangaroos were killed over 20 years (1960 to 1980).

A classical example of over-exploitation and of over-hunting is the fate of Steller's sea cow (*Rhetina stelleri*) which had been exterminated by traders in the Commander Islands only 27 years after its discovery. Immoderate trade has brought many species of cetaceans such as large baleen whales (Balaenidae), humpback whales (Megaptera), grey whales (Eshrichtius) and blue whales (*Balaenoptera musculus*), to the verge of extermination.

The Convention on International Trade in Endangered Species of 1973 (CITES) should tentatively become a bar to the mass extermination of rare species: by 1985 it had been signed by 88 countries. However, high prices for animals and their trade products are too great an incentive for illegal trade. In 1979 the US Fish and Wildlife Service reported an illegal organization which within only 18 months had sold about 46,000 hides of South American wild cats, 30,000 ocelot hides, more 15,000 otter specimens, almost 6000 leopards, 1900 cheetah hides, 500 cougar hides and 200 jaguar hides, estimated at US $5 million. In 1984, after a 3-year investigation in the USA revealed another equally large illegal organization. In November of 1980 CITES reported illegal trade in disappearing species amounting to US $12 million. One merchant in Frankfurt (FRG) received, under false documents, 200,000 cayman hides, 40,000 ocelot hides, 140,000 otter hides and hides of other animals from Paraguay. These commodities had been transported illegally to Paraguay from Brazil, Argentina and Bolivia. Similar methods with forged documents by fur suppliers have been exposed in Denmark, Switzerland and the USA. Big international swindles with forgeries of CITES documents are revealed every year.

In late 1979 at the border between Mexico and the USA illegal operators were arrested who had transported a batch of 17,500 hides of lynx, jungle cat and silver fox for sale on the West German market. At least 300 thousand llama hides have been brought illegally into the FRG from South America over the past years. Thus, in 1979 alone and only from Paraguay some 230,000 ocelot hides were exported into the FRG. The European market receives annually a total of about 700,000 wild cat hides of medium and small size. In 1980 a coat of ten ocelot hides reached a price of US $40,000 in the FRG, i.e. the same as the cost of three cars of the luxury class. In Bangkok (Thailand) the price for the claws and teeth of only one tiger was as much as US $4000 in 1983.

Illegal procurement is one of the principal causes of decrease in the number of many species of large mammals. The turnover in the ivory trade increased from 400 tons per year in 1968 to 10,000 tons per year in 1982, which had meant killing more than 100,000 elephants in Asian and African countries.

Procuring elephants is performed mainly illegally by hunters sometimes equipped as military troops, with airplanes and helicopters, machine-guns and poisons. In 1979 the price for ivory was more than US $100/1 kg on the world market.

The results of illegal extermination of rhinoceros, which is prohibited for trade in all countries, are more dangerous: in 1979 alone in Bremen and Frankfurt (FRG), customs discovered 220 freshly sawn horns imported from Kenya. In 1960 about 16,000–20,000 rhinoceroses inhabited Kenya, by 1979 their number had decreased to less than 1500 specimens. In 1976 in Tanzania there were still 2500 rhinoceroses, and in 1983 their number was only 50. The price of rhinoceros horn per 1 kg rose several tenfold during this time and reached US $11,000 in 1980. The horn comminuted in powder and prepared as a drug costs about US $600 per ounce in Japan and Singapore. As a result of massive illegal trading across all the African continent, the number of rhiniceroses has decreased tenfold.

The scale of catching animals for keeping in captivity is significant. Thus, some 200,000 monkeys were kept in houses and personal collections alone in the FRG in 1980. It should not be forgotten that to catch small monkeys, about ten specimens are ruined for one animal caught, and a further 50% of the animals are lost during transportation. However, all costs of these predatory operations are returned since one specimen of a rare monkey (e.g. a pygmy king-kong), for example, have a price of up to US $5,000 on the USA, and in 1984, young gorillas cost up to US $70,000. Some 200 photographers at Spanish health resorts use young chimpanzees to be photographed with tourists. To maintain this business about 800 to 1500 animals are procured every year from the wild.

The scale of using animals for testing purposes is significant. In the USA more than 1.5 million macaques were used to produce the polyomyelitis vaccine between 1954 and 1960; up to 200,000 macaques were brought into the USA annually from India during late 1950's. In 1977 and 1978 the import of primates into this country was 28,559 and 22,630 specimens, respectively (58% from Latin America, 34% from Asia and 7% from Africa). In Peru about 7.5 million monkeys were killed between 1964 and 1974 for use as food and 1.5 million were killed for export; at least 200,000 primates were sold annually in the late 1960's.

It should be mentioned that WHO, together with IUCN, made a decision in 1983, that the use of rare, disappearing or endangered primate species for medical purposes is permitted only when animals are taken not from the wild, but from propagating colonies which live in captivity.

4.2.2.2 Birds

The scale of the extermination of birds both in the recent past and at present is enormous. The destruction of birds as a result of direct trade or catching for collections is shown in the data in Table 4.13. At wintering locations in Southern Europe, about 80 to 90% of the birds wintering there appear to be shot.

Devastation among birds of prey which can be trained for hunting is stimulated by the very high prices resulting from the great popularity of falcon hunting in Arabic countries. A falcon fledgling procured in Island is estimated at 2 million Islandic krones; at the bird market in Bombay, a pair of falcons was sold for 15,000 rupees, etc. The price for some birds of prey exceeds US $45,000. It was

Table 4.13. Volume of trade or catching of some birds. (Yablokov and Ostroumov 1983; Inskipp and Wells 1979 and others)

Group of species	Region	Number of birds procured annually	Notes
All groups, living birds	All the world	7 million	Volume of trade for aviaries, zoos, keeping at home, 1975
All groups of small singing birds	USSR	Several hundreds of thousands	Calculated data, the late 1970's
All parrots	All the world	About 1 million	Catching for keeping at home, the late 1970's
All groups of wild birds	USA	300 thousand in 1976, 422 thousand in 1979, 778 thousand in 1982	During 1976–1984 the USA legally imported about 5 million wild birds, of which about 2 million birds were lost during transportation or quarantine

not surprising that in Central Spain 59 falcon nests out of a total of 79 which were under observation and protection were robbed in 1983. Prices for contraband Amazona parrots to be sold in the USA are very high; e.g. a hyacinth mohao had a price of up to US $5000 in 1980 (with a record price of US $15,700).

The extermination of small passeriform birds has acquired a really inconceivable scale in Europe and Japan. In India alone during the bird migration period, about 180 million birds are procured, including about 80 million birds from ambushes, 40 million birds by nets, 30 million birds by means of glue. In France by means of nets alone up to 500,000 larks and about 250,000 greenfinches, linnets, chaffinches and yellow buntlings are caught annually, the scale of this procurement being continuously increased (Table 4.14).

According to our calculations, in the markets of large towns of the European part of the USSR at least several hundred thousands of small singing birds are sold annually. The official international trade with wild birds reached 7 million specimens in 1975 (Fig. 4.3). It is calculated that 80 to 98% of the wild birds transported over long distances are lost during transportation or shortly after arrival (Inskipp and Wells 1979). Thus, for example, of 778,000 wild birds

Table 4.14. Volume of annual procurement of some bird species (or species groups). (Dezhkin 1983; Priklonsky 1977; Anderson 1985; and others)

Species (group of species)	Country, region	Number of birds procured	Notes
Emu (*Dromiceus novaehollandiae*	Australia	More than 100,000	Extermination to protect agricultural production, 1932–1935
Mourning dove (*Zenaida macroura*)	USA	More than 2.5 million	Hunting, 1970's
Thrushes	Majorca	3.25 mill.	Hunting, late 1970's
Starling	Greece	2.9 million	Hunting, 1970's
Singing birds	Japan	4 million	Procuring for restaurants, early 1980's
Bog game (ducks, geese, snipes, etc.)	Europe	16.59 million	1970's
Field feathered game Pheasants, partridges, quails	Europe	29.49 million	1970's
Forest feathered game (four species of grouse, doves)	Europe	9.52 million	1970's
All migrating and wintering small birds	Southern Europe	Up to 300 million (in Italy 150–200 mill., France, 40 million)	Procuring by all methods, late 1970's
All hunting and trade birds	USSR	About 50 million	Hunting, 1971–1975

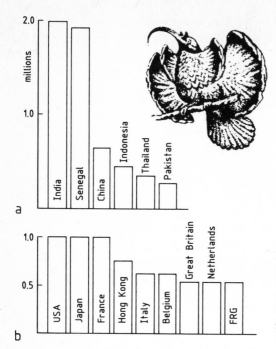

Fig. 4.3a,b. Level of official export (**a**) and import (**b**) of non-domestic fowl by major countries involved in the international fowl trade. The highest annual figures for the period of 1970–1975 are given for each country. (Inskipp and Wells 1979)

officially imported into the USA in 1982 about 56,000 specimens were found to be dead on arrival in the country and 147.8000 were destroyed during veterinary quarantine. In the FRG about 7 million wild birds were held in captivity in 1984.

One serious threat for the bird population is a recent fashion to collect stuffed birds or bird's eggs. For example, in 1985 it have been reported that in Northern Sweden, two illegal traders from West Germany had in their car 191 corpses of tawny owls, *Strix nebulosa*. In Australia a collection of bird's eggs comprising more than 33,000 specimens including eggs of 500 disappearing species of birds has been detected and confiscated. In 1986 in Austria a collection of 35,000 eggs illegally collected in Finland were confiscated.

4.2.2.3 Reptiles and Amphibians

The scale of their extermination is enormous and usually underestimated even by zoologists. In the course of 300 years (17th-19th centuries), whale traders and sailors exterminated about 10 million Galapagos tortoises on the Galapagos Islands. However, the scale of recent extermination of reptiles and amphibians is no less vast. In 1980 alone, more than 5 million crocodiles of various species were exterminated for hide processing. In 1976, Indonesia officially exported 270,000 hides of monitor lizards, 28,000 crocodiles, 350,000 snakes, 71,000 turtle shells and about 50 million frogs.

The annual volume of sales of Mediterranean turtles in zoological shops in European countries approached 250,000 specimens in 1938 and maintained a level of about 100,000 specimens per year for a long time. In the late 1970's in the FRG about 1 million tortoises of various species were being kept as pets, and only 10 to 20% of the animals survived for more than 2 years. Exporting has devastated the tortoise population of the species *Testudo graeca*, *T. hermanni*, *T. horsfieldi* in Spain, Morocco, Turkey, Yugoslavia, Greece and the Italian islands. For the tortoise, *T. marginata*, which was always rare, the state was critical. In England alone about 2 million tortoises were imported from 1965–1975, and a total of about 10 million small live tortoises were sold throughout the world during this decade.

Large edible turtles have been under strong pressure: all seven species are under threatened with disappearance. Rates of decrease in the number of certain species are catastrophic: about 40,000 adult females of the Atlantic turtle (*Lepidochelis kempii*) were counted along the shore from the Mexican Gulf to Massachusetts in 1947, only 1500 1966; in 1976 only 256 females were counted, and only one single beach was under protection.

According to computations based on the underestimated official data on the volume of trade with shells of large turtles, 170,000 to 240,000 turtles were sold throughout the world, 100,000 to 140,000 in 1977 and up to 300,000 in the 1980's. The larger proportion of animals (about 60%) were supplied from Southeastern Asia and from the Pacific Basin, 15% from islands of the Caribbean Basin and 10% from the Central American countries. Altogether turtles have been procured in almost ten countries.

The following facts suggest the range of the procurement of snakes. In 1985 alone 1.3 million live snakes were exported from Thailand, mainly for restaurants in Hong Kong, South Korea, Japan and the USA. In 1983 alone about 335,000 snake hides intended for illegal export were confiscated in India; one single contraband batch detained in the port of Cochin consisted of 60,000 snake hides.

The scale of crocodile extermination is also very great. The export of 53,433 hides of the black cayman (*Melanosuchus niger*) and 10,164 hides of the crocodile cayman (*Caiman crocodilus*) was recorded from the Peruvian part of the Amazonas Basin between 1962 and 1969. In 1974 alone about 556,422 hides of the latter species were exported from this region. About 1.5 million crocodile hides officially entered the world market in the 1980's, but at least a further 1 million hides entered illegally (mainly from South America). One contraband cargo alone, which was found in a Greek ship in the port of Rio de Janeiro in 1985 contained the hides of 45,000 animals.

The largest volume of processing of raw leather materials of reptiles appears to be in Japan. According to the official customs data, in 1978 Japanese import of such hides was as follows: 56 tons of lizard hides from Bangladesh, 22 tons from Singapore, 19 tons from Indonesia, 14 tons from Pakistan; 6.2 tons of snake hides from Indonesia, 4.5 tons from the Philippines, 3 tons from Thailand. About 103 tons of hides of alligators and crocodiles (including 78 tons of cayman hides from Latin America, among them 54% from Paraguay, 20% from Columbia) have been imported for the same year.

In addition to Japan, major importers of reptile hides for the past years have been the USA, Great Britain, Singapore and France. A Singapore-made handbag of crocodile leather cost more than US $1000 in London in 1978 and one processed crocodile hide of 12″ to 14″ in width cost about US $200. One London-based company alone, Mappin and Webb, sold about 10,000 handbags of lizard leather annually for several years till 1978 (to make one bag 6 to 12 hides of large lizards are required).

Calculated data on the annual procurement of the principal groups of reptiles and amphibians are presented in Table 4.15.

Table 4.15. Calculated data on the annual world trade of principal groups of reptiles and amphibians for the early 1980's

Group	Volume of trade	Applications
Largest turtles	0.5 to 1.0 million specimens	To produce meat, shells, raw leather materials
Small land or freshwater tortoises	At least 1.5 to 2.0 million specimens	To hold in captivity
Crocodiles	2.0 to 3.0 million specimens	As raw leather materials
Snakes	3.0 to 5.0 million	To produce meat, poisons, to hold in captivity, as raw leather materials
Small and medium lizards	Tens of millions specimens	To produce meat, for hide processing, to hold in captivity
Monitor lizards (three species)	2 to 3 million specimens	
Frogs	250 to 300 million specimens	Mainly for producing meat

4.2.2.4 Fishes

The number of fishes procured is difficult to assess. Calculated data has shown that at least 19 billion specimens of codfish were procured by all countries during the period of intense fishing of this species in the North Atlantic. The procurement of other commercial fish species also appears to be estimated at billions of specimens.

The Philippines alone sold annually in the late 1970's about 2 to 3.5 million live tropic fish caught in the wild to be kept in aquariums, and the export of live fish from Brazil reached 12 million specimens in 1982. About 40 million aquarium fish were sold through Singapore annually in the mid-1970's; in 1977 alone about 100 million tropical fish were exported officially into the USA. In the mid-1980's in the FRG, about 50 million fish were being kept in aquariums. A total of at least 2,000,000 species of aquarium fish are involved in the trade.

4.2.2.5 Invertebrata

The scale of insect catching is also significant. From Taiwan alone up to 500 million various butterflies are exported annually for sale (at a cost of about US $30 million); Brazil sold up to 50 million specimens of the male butterfly *Morpho* sp. alone per year in the mid-1960's. The price for certain rare specimens of the *Ornitoptera* genus was US $2000 in the late 1970's. The butterfly species *Strymonida iyonis* which was found on Shikoku Island in 1957 was almost completely exterminated by collecters.

In the USSR, the massive collecting wrought serious damage to the populations of rare species of large butterflies of the European part of the USSR, of the Caucasus, and to rare beetles in the far east of the USSR.

Calculations have shown that trade catching alone of crayfish in the basin of the Caspian Sea and in the Central European part of the USSR exceeds 90 million specimens per year (Yablokov and Ostroumov 1983).

Excessive procurement endangers many other species, especially mollusks. Thus, the Philippines alone exported about 3500 tons of decorative shells in 1979. The import of shells in the USA increased from about 1000 tons in 1969 to 4400 tons in 1978. The requirements for unprocessed shells *Trochus* (source of nacre) in Japan, Taiwan and South Korea constitute about 6000 tons per year (about 5 to 8 million specimens are procured). Excessive harvesting of giant shells of the Tridacnidae family has resulted in the disappearance of tridacnes in many coral reefs of the Pacific. *Tridacna derasa* and *T. gigas* have disappeared in certain parts of West Indonesia. A study of the populations of tridacnes in the Philippines has shown that, within the sea reservation, the total biomass of four species (*T. hippopus, T. crocea, T. squamosa, T. maxima*) amounted to 79 to 260 kg/ha, and beyond the borders of the reservation it is two- to sixfold less. In 1983 alone more than 20 million edible snails were exported from Hungary.

Some large spiders have become popular pets. Thus, for example the tarantula, *Brachypelma smithi*, which is rare in nature and lives in the Mexican semi-arid lands, has attracted attention by its beauty, easy maintenance and long life (up to 20 years) in home conditions. The volume of its official export from Mexico approached 20,000 specimens in 1984 (volume of illegal export being several times more). The procuring of such a rare species in these amounts threatens its existence.

4.2.2.6 Plants

The share of plant species which are disappearing due to excessive gathering because of their decorative, technical, medicinal or nutritional properties is large. Thus, collecting cactuses has became of increasing importance. For example, about 260,000 cactuses were taken from deserts in Arizona (USA) in 1977. The owner of a small botanical garden in Texas (USA) was sentenced to jail for one year in 1983 for the forgery of documents permitting the import of 100,000 cactuses to the United States. About 3600 pieces of cactuses of various species delivered illegally from Mexico were confiscated from a tourist group of Frankfurt, FRG, in 1978. Although in the total world cactus trade only 15 to

20% pieces are taken from nature and the remaining portion is cultivated by artificial propagation, these facts should not create the illusion of satisfaction. With the enormous scale of the trade, these 15 to 20% are enough to endanger many natural populations. The state of populations of rare or recently described cactuses is particularly alarming. Frequently in economic terms the trade with plants delivered from natural habitats is more profitable if they are near extermination. To save cactuses from collecters, the authors of publications describing new species frequently avoid disclosing the exact location of their growth.

In recent years the collecters' interest in other succulents, such as Pachipodium, Euphorbia, Anacampseros and aloe, has risen significantly, and the total volume of their trade is ahead of cactuses.

Many other groups of plants (not only succulents), and particularly the family of orchids have long suffered from unlimited gathering for collection and trade. For example, about 17 tons of ginseng, *Panax guinguefolius*, were exported from the United States in 1977, only 200 kg of it legally. This resulted in the total extermination of populations of this species in many regions of the Appalachians. The golden slipper orchid, *Paphiopedilum urmeniacum*, discovered in the Yunnan Province (China) in 1982 apparently had disappeared in nature by 1985 because of collecters, who paid about £ 150 per individual flower. This species has shared the fate of a closely related species of orchid, *P. delenatii*, which since 1924 has existed only in collections.

The scale of international trade with rare plant species is characterized by the following figures: annual imports of plants covered by SITES into the USA alone appears to be about 10 million pieces (Threatened Plants Newsletter 1983). The requirements of SITES in the control of world plant trade are meeting with little success.

A dangerous type of excessive use and procurement of plant resources is lumbering in species-rich ecosystems, primarily in tropical forests. Worldwide procurement of hardwood in the tropics increased fourfold between 1950 and 1983. According to predictions the laying-in of tropical hardwood will rise threefold between 1973 and 2000 and in America even sixfold for the same period. The expected demand in tropical wood in 2000 will be so great that by that time the forest ecosystems will be exterminated in 59% of 935 million ha of the tropical rain forests if the world that remained on the Earth in 1973, and which, in their turn, were only 58% of the initial area of the wet tropical forest, which occupied about 1600 million ha in the past (Soule and Wilcox 1980, Chap. 17).

Excessive procurement is the most destructive factor for rare or disappearing plants of the USSR flora (see Table 4.11). According to calculated data, about 42% of these species suffered from this. At least 87 rare or disappearing species of USSR flora have suffered extermination for their decorative properties. Among them, in particular, are many representatives of such families as orchids, peony, lily and others.

At least 39 rare species or subspecies have become endangered, in particular by procurement for technical needs, mainly for obtaining wood. The scale of plant extermination for producing medicinal raw materials and food products is hardly less, 30 and 29 species, respectively. All these figures concern only rare or disappearing species of the USSR flora. The total number of all species of our flora whose populations are suffering from this factor is much greater. The above figures as a whole are slightly unexpected, since the results of death cause analysis which was made for vertebrates were previously applied automatically to plants, and it was assumed that the factor of habitat destruction was for them also, the most important.

4.2.3 The Influence of Introduced Species

This is the primary cause invertebrates for species already exterminated, and holds third rank for species approaching extermination.

The influence of a newcomer, a possible result of which is the extermination of the local species, takes four main directions: absorption of closely related forms as a result of hybridization; competition for space or food; direct persecution and destruction of habitual biotopes (Table 4.16). The influence of introduced species is most evident on the fauna of islands and hydrobionts (Fig. 4.4), where the extent of introduction is high. More than 2500 species of insects have penetrated North America from other continents, and the overwhelming part of them has been introduced inadvertently by human action — with luggage, goods, cloth, nursery plants, etc. For example, in the water area of the San Francisco Gulf were found some 200 invertebrate species and more than 20 fish species which are newcomers by intention or chance (Nichols et al. 1986).

In Czechoslovakia, 25 fish species were introduced or expanded the area of their distribution due to anthropogenic conditions as against 42 species of endemic (local) fishes. During the last 100 years, 12 species have disappeared from the ČSSR ichtiofauna and 18 endemic species (i.e. 43% of all endemics) are threatened as a result of anthropogenic changes. Among the threatened species are *Acipenser ruthenus*, *Umbra krameri*, *Pelecus cultratus*, *Gabio uranoscopus* and *G. kessleri*.

On the Hawaiian Islands up to 1978 were introduced 22 mammal species, about 160 bird species, about 1300 insect species and more than 2000 flowering plants. Mainly for this reason, during the period since the discovery of these islands to the present, at least 22 bird species (30% of the aboriginal ornithofauna) have become extinct and 70% species of the Hawaiian flora are under threat of extinction (Berger 1979).

One hundred percent species of terrestrial mammals (29) and about 20% of the bird species (31) of New Zealand have been introduced. In Great Britain over the last decades have been acclimatized three deer species (muntjac, sika, Chinese water deer), wallaby, several species of ducks, geese, pheasants, edible dormouse, two species of frogs, etc. In Poland there are 21 species of introduced mammals, birds and fishes, which represents about 5% of the total number of vertebrate species propagating in this territory.

Table 4.16. Some recent examples of the negative effect of introduced species. (Data from various authors)

Species, region of the origin	Region of the introduction, time	Causes of introduction	Consequences
Tamarix (*Tamarix pentadra*)	North America, 1930's	To control soil erosion	Has propagated across all the Western states of the USA, is displacing the local species
Predator mollusk *Englandina rosea*	Some islands of the Pacific Ocean, 1950's	To decrease the number of giant African snail *Achatina fulica* introduced earlier	Extinction of many endemic species of terrestrial mollusks
Nile perch *Mormyrops* sp.	Lake Victoria, late 1950's	To supplement food resources	Extinction of many aboriginal species
Piraya *Rooseveltiella nattereri*, South America	Malaysia, 1980's	Occasional settlement?	Decrease of the number of many local species
Giant frog, South America	Queensland State, Australia, 1930's	To protect sugar cane	Has propagated widely across the country, displacing the local species of amphibians; extermination of some insects (including bees)
American mink (*Mustela vision*)	Europe, 1930's	Escapes from nurseries and to enrich the fauna	Displacement of two muskrat species, otters, European mink, extermination of water fowl
Mongoose, India	Caribbean countries, 1872	To control rodents	Extermination of many local species of small vertebrata, damage to agriculture

Introducing parasites or disease agents is especially dangerous, since the local species frequently have no protection against them. Thus the fungus *Endothia parasitica*, brought to the USA from Asia with nursery plants, has almost completely exterminated the Eastern American species of chestnut *Castania dentata*. Disease agents or parasites which are inherent but harmless for the newcomer can be fatal for the aboriginal species. An attempt to acclimatize Caspian starred sturgeon in the Aral basin is one such example. This attempt failed, but the gill fluke *Nitzschia sturionis* that was brought in concurrently resulted in a massive loss of endemic Aral sturgeon *Acipenser nudiventris* (Nikolayev 1980).

The introduction of goats on the Island of Saint Helena has resulted in the total extermination of all 33 endemic plant species. More than 900 foreign species

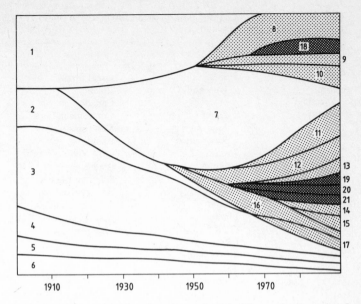

Fig. 4.4. Example of the negative effect of introduced species on aboriginal fauna: the dynamics of change in the ichthyofauna of Lake Balkhash in Kazakhstan (Central Asia). *1–6* aboriginal species (*Perca schrenki, Schizothorax pseudoksaiensis, Schizothorax argentatus, Nemachilus strauchi, Nemachilus labiatus, Phoxinus poljakowi*); *7–17* specially introduced species (*7 Cyprinus carpio; 8 Lucioperca lucioperca; 9 Aspius aspius; 10 Siluris glanis; 11 Rutilus rutilus; 12 Abramis brama; 13 Leuciscus leuciscus; 14 Ctenopharyngodon idella; 15 Tinca tinca; 16 Acipenser nudiventris; 17 Barbus brachycephalus*); *18–21* accidentally introduced species (*18 Lucioperca volgensis; 19 Pseudorasbora parva; 20 Pseudogobio rivularis; 21 Percottus glehni*). Additionally, at least 12 invertebrate species (mollusks, polychaete, crustacea) have been introduced. The figure shows the relative abundance of various fish species in various years. (Based on various sources)

are found in Madagascar flora, more than 700 such species in British flora, about 200 of a total of 1100 species in the Karelian flora have been introduced by human agency. Summary data on the anthropochorous species in spontaneous floras of some regions are presented in Table 4.17.

There is evidence that introduced plants can possess increased ability to suppress the local species due to the newcomer's secretion of allelopathic substances which are unusual for the aboriginal species. This has been shown, for example, for hawkweed (*Hieracium pilosella*) introduced in New Zealand (Ostroumov 1986).

Formerly, the wide settlement of the tropical freshwater fish *Gambusia affinis*, that exterminates mosquito larvae, was considered as a positive example of introduction.

In some cases the secondary disturbance of the habitats of other species, and even the action on the abiotic component of the ecosystems prove to be dangerous results of the introduction. Thus, the settlement of red deer in New Zealand has resulted in the elimination of the undergrowth, and as the consequence in the

Table 4.17. Share of anthropochorous species of vascular plants in some spontaneous floras of the world. (According to various authors)

1 to 10%	10.1 to 20%	20.1 to 30%	more 40%
Hungary	USSR	Great Britain	Region of
USSR	Kiev and	Ireland	San Francisco
Yaroslavl	environs	Canada	City, USA
district	Karelia	Connecticut, USA	Mocha Island,
Crimea	Lapland	São-Tome	Chile
Estonia	Reservation	Islands	Juan Fernandez
Belorussia	Austria	Cape Verde	Islands (Chile)
Udmurtia	Belgium	Islands	Kerguelen
Kuril	Netherlands	Solomon Islands	Island
Islands	Norway	New Zealand	Seychelles
Amur	Poland		Tahiti
Region	Finland		Region of
Primorsky	GDR		Szczecin City,
Territory	FRG		Poland
Yugoslavia	Switzerland		
Madagascar	Sweden		
	Pakistan		
	USA		
	Texas, USA		
	Hawaii		

decrease of the number of the non-flying owl parrot (*Stringops habroptilus*), the only representative of the owl parrot genus. The settlement of hare *Lepus capensis* in the Hawaiian Islands has resulted in the extinction of several species of butterflies of the Noctuidea family. The propagation of the water hyacinth (*Eichhornia crassipes*), which during transpiration can increase the evaporation of water more than seven times in comparison with the open surface of the aquatic reservoir, may change the total hydrological regime of even large aquatic ecosystems.

Sometimes undesirable assimilative hybridization between related species is possible as a result of disturbances in ecosystems induced by the new comers. This phenomenon has occurred in the case of certain endemic freshwater fish of the Cape Province.

Of seven foreign species of the game animals which have been introduced for hunting in the USSR in the 20th century six species affect the local fauna negatively.

4.2.4 Special Extermination

Until now many animals have been exterminated to protect agricultural products or trade objects. For example, fishers exterminate *Pinnipedia* animals or otters, claiming protection of the fish yield as a pretence; large cats and predator birds have been exterminated under the pretence of protecting agricultural animals.

In Australia kangaroos were exterminated till 1917 as agricultural pests: in Queensland alone some 26 million grey kangaroos and red kangaroos were killed over 40 years. As a further example of animals which had been killed under the pretence of protecting agricultural production are representatives of the Australian marsupial family of wombats. Initially, they were exterminated for damaging pastures with their holes. Later, they became victims of campaigns for the extermination of rabbits. In this manner the hairy-nosed wombat (*Lasiorhinus latifrons*) had been completely exterminated in the Province of Victoria. Populations were to be found at the beginning of the 1970's only in the narrow shoreline zone of Southern Australia.

About 900 griffon vultures were killed from 1978 till 1982 in the French Savannas, and about 800 eagles were killed by the pilot of one single airplane in Colorado and Wyoming in 1971 alone.

In France animals considered as "harmful" until 1983 included weasel, wild cat, marten, ermine, pole cat, starling, magpie, jay, wild doves, and owners or leasers of land plots or areas had the right to kill them without shooting permission.

The presentation of widespread measures for controlling "pests" or "weeds" has resulted in certain species of animals or plants becoming endangered by complete extermination. Some "weeds" or ruderal plant species are included now in the Red Data Books of a number of countries.

The experience of many countries has shown that estimates of the damage inflicted by predators on agricultural animals or by wild Ungulata on cultivated crops is usually exaggerated. Thus, the damage inflicted by coyotes on sheep herds in North America is much less than the cost of controlling them. Accusing saigas or water voles of supposedly exterminating crops in Northern Kazakhstan also appears to have been considerably exaggerated.

4.2.5 Occasional (Unintended) Extermination, Incidental Kills

Many animals may be killed as a result of natural catastrophes. For example, the eruption of the Saint Helens volcano (USA) in 1979 was the cause of death of more than 5000 sand moose, 6000 deer, 200 bears, 100 goats and 15 cougars. However, the scala of anthropogenic perishing are much greater.

4.2.5.1 In Oceans or Interior Aquatic Reservoirs

Occasional elimination is often a result of non-selective trading by means of nets of various types. At present, the extent of fishing is such that co-elimination may pose a threat of dramatic decrease in the number of species which were widespread and very numerous before. For example, the annual capture of fish by means of shrimp trawl nets approaches 21 million tons (that equals the amount of fish consumed in all the developed countries), i.e. billions of specimens of many fish species are perishing this way every year.

About 1200 dead sea turtles thrown away by fishers were found on the eastern shore of the USA over a period of only 2 months in 1980. The cause of their death has been determined as setting caught in trawl nets during shrimp fishing. In the process of tuna fishing in the Central Pacific up to 400,000 dolphins of *Stenella* genus perished per year in the mid-1970's. The scale of this extermination has diminished by about 20 times as a result of the widespread use of special nets which allow dolphins to escape; nevertheless the number of dolphins that die remains very high. Many dolphins, seals and birds die in fishing nets in the Northern part of the Pacific Ocean. Thus, in the North Atlantic up to 500,000 thick-billed guillemots (*Uria lomvia*) perished per year from 1965–1975; up to 17% of their population died annually in nets off the shore of California in the beginning of the 1980's; and more than 6.5 million guillemots, puffins, Atlantic puffins, albatrosses, seagulls, and fulmars perished in the nets of Japanese fishers in the North Pacific from 1952 to 1978. Floating net patches pose an increasing danger. At least 50,000 seals died per year in the world oceans during the 1980's.

Many sea animals are killed in nets made to obstruct sharks. In Queensland (Australia) alone, about 20,500 sharks, 468 dugongs, 317 dolphins, 2654 turtles and 10,900 skates died in such nets between 1965 and 1980.

The loss of aquatic organisms as a result of operating ships is also high. Direct estimates have shown that one single tanker exceeding the permitted speed near one of the islands within the Volga river delta has resulted in about 126 thousand fries of various fish species being washed onto the shore and dying.

The mortality of various hydrobionts at water intake constructions is enormous. Thus, about 7 billion fingerlings died in the irrigation systems of Uzbekistan in 1966. At the Lower Volga in the beginning of the 1970's the loss of fish fingerlings at the water intake plants was close to 6 billion specimens (three times higher than the capacity of all the fish hatcheries). About 400 million fingerlings perished in water intake plants in the North Crimean Canal during 2 months in 1971. Up to 6% of the fishing potential has been entrapped and died at water intake grates of 14 thermal power plants and six nuclear power plants located near the shores of Lake Michigan (USA).

4.2.5.2 On Highways

Vertebrates or invertebrates running or flying across the highways are also often killed by moving vehicles. The resultant loss in fauna is high. As many as 1.6 million insects are crushed on the windshield or hood of one single motor vehicle covering about 10,000 kilometres per summer season. About 140 deer per year perished on the US highways in the mid-1970's, and the total number of vertebrates killed on highways of that country in 1970 alone amounted to 365 million specimens (including small rodents, reptiles, amphibians, birds). In Denmark, about 10 million various vertebrates (including 6.3 million frogs, more than 100,000 hedgehogs and about 120,000 hares) were killed annually on the highways during the 1960's. About 80,000 roe deer and 120,000 hares, 2000 boars, 1500 fallow deer and 1000 deer perished annually on the highways in the FRG during the beginning of the 1980's. In Slovakia the loss of quails on the highways is only half of the number that is shot in the hunting season, up to 23% pheasants and 14% hares compared with the number of animals shot. In Australia at the beginning of the 1980's about 20 to 30% of the annual offspring of hares (*Lepus europaeus*) were killed under wheels of motor vehicles, and in Holland about 10 to 15% of all roe deer (*Capreolus capreolus*) have perished on the highways. About 4.9% of all the Ungulata that died in the Berezinsk Reservation (Belorussia) were killed on the roads as a result of collisions with motor vehicles. The proportion of animals dying is higher in more populated countries: in the GDR, the share of roe deer killed as a result of collisions with motor vehicles is about 24%; in the FRG about 15% of the fallow deer die; in a number of districts in France, up to 25.6% of the hares are killed on the highways. Extrapolating data available for certain districts of the USSR assume that more than 10,000 moose and deer and apparently no fewer roe deer and boars die annually on the highways in the USSR (Yablokov and Ostroumov 1983).

Calculations have shown that even at a traffic intensity of only ten motor vehicles per hour, up to 30% of all toads crossing the highways have been run over. An average of about ten birds per kilometer die on the roads of the steppe zone of the European part of the USSR during the three summer months. For the remaining months of the year this number increases to at least 15 specimens. Our estimates suggest that at least 9 million birds have perished annually on the highways of the country in the late 1970's.

The extent of bird loss on the highways of Western Europe is even more extensive: up to 170 specimens per 10 km per month for July-September and 20 and 50 specimens per month for the rest of the year. It seems that more than 30 million birds die annually on the European highways (without the USSR).

Animals (mainly birds) are also killed on collision with flying planes or under the action of sound waves generated by modern planes or spacecraft. A significant threat for birds nesting near airports appears to be the low frequency noise generated on breaking the sound barrier. This results in decreased of clutches and fledgling hatching rate, increase of aggressivity and mortality in colonies of silver seagulls (*Larus argentatus*) (Burger 1981).

Finally, the increasing development of icebreaking operations also appears to cause the loss of many animals. Thus, a significant number of Arctic deer (*Rangifer tarandus*) perishes when crossing from the Arctic islands to the mainland in air holes left by powerful ice breakers. On the Svir River and in other parts of the Belomor-Baltic River System, the artificial prolongation of the navigation period by means of ice breakers also results in the loss of appreciable numbers of wild Ungulata such as moose (*Alces alces*).

4.2.5.3 In the Fields
The occasional loss of animals in agricultural production is also enormous. Thoughtless agricultural technology and the careless use or unsuitable storage and application of pesticides and other farm chemicals are the main causes.

Until now, when mowing grasses or harvesting fields, agricultural machines usually run from the periphery to the central part of the field, where frightened animals concentrate and then die in the machines. In a number of districts of the USSR more hares, partridges and quails die in this way than are shot by hunters.

About 46% of the roe deer perished in some districts of GDR; more than 29% of the hares and 18.5% of the partridges that died in some districts of France were killed by agricultural machines. Up to 7.7% of the newborn roe deer in Poland are killed by agricultural machines. In Czechoslovakia, during mechanical harvesting of the fields, ten times more animals and birds perish than hunters procure.

Since it is impossible to isolate agrocenoses where pesticides or herbicides are applied from the adjacent ecosystems, these substances produce serious negative effects on the number of many non-target species of animals or plants. About 11.2% of the hares of the total number of those killed in 1976 to 1980 in certain districts of France died as a result of plant protection chemicals. An important side effect of pesticides consists in the abrupt decrease in the species variability of ecosystems. Another pesticide property which is dangerous for rare or disappearing forms of animals is that pesticides affect individual specimens, regardless of species density.

Random examination in the USSR has shown that of all the causes of occasional death of mammals associated with agricultural production, more than 41% lethal cases are due to poisoning with mineral fertilizers, about 22.4% with herbicides and about 16.7% with zoocides. Similar figures for birds (forest game, grey partridges, bustards, ducks, geese) are 13.7, 19.2 and 57.9%, respectively. Numerous moose perish due to poisoning with strewn fertilizers such as urea, potassium nitrate and ammonium nitrate. The almost complete disappearance of the butterfly *Euproserpinus weisti* is an example of the dangerous consequences of using pesticides in agriculture. The last habitat of this butterfly in Colorado (USA) was treated with Malathione to control locusts in 1980. As a result, only 200 caterpillars and about 25 adult insects had survived by 1981. A further example is the extinction of the British population of Black-veined White (*Aporia crataegi*) as a result of treating fields with insecticides.

The agricultural burning of dry vegetation represents a real disaster. These operations kill numerous small and large birds, mammals, and result in the destruction of habitats suitable for nesting.

4.2.5.4 Mortality Associated with Petroleum Production

The contemporary complex of petroleum-related industries and the production and transportation of petroleum products can be a cause of the occasional killing of migratory birds. At one gas flare alone where the gas of the crude oil production in the North Sea is burned, several tens of thousands of migratory birds (starlings, thrushes, herons, oystercatchers, etc.) died during 5 days in October 1976. In only one night, October 25, 1975, at least 3000 birds died in flares at one of the crude oil-producing rigs in the North Sea. Many millions of birds perish annually in gas flares, thousands of which are burning now in the crude oil-producing regions around the world.

A large number of birds also die in oil spills in the ocean and on the mainland when crude oil is stored in open reservoirs, leading to the contamination of feathers with crude oil. Calculations have shown that hundreds of thousands of birds die annually in the oil spills in the North Atlantic alone.

4.2.5.5 Mortality Associated with Power Transmission Lines

The killing of large birds, e.g. tawny eagles, imperial eagles, golden eagles, harrier eagles, sakers in the USSR and red kites and white storks in the FRG on wires or masts of high-voltage power transmission lines has become more frequent over the last decades. The mortality rate of tawny eagles in Kazakhstan, Uzbekistan, Kalmykia, and other territories of the steppe zone of the USSR is especially high (up to 35 birds per 10 km annually) (Peskov 1982). Calculations show that the total may be hundreds of thousands of specimens dying per year. Collision with high-voltage power transmission lines was the cause for 75% of deaths of young storks in France, the FRG, and Switzerland at the beginning of the 1980's.

4.2.5.6 Mortality as a Result of Environmental Pollution

Pollution of the waters of the Rhine (FRG) with the insecticide Endosulfane on June 23, 1969, when only one drum of this chemical was lost, is a striking example of the tragic consequences of incorrect storage of plant protection chemicals. As a result, up to 20 million specimens of fishes died over a distance of several tens of kilometres within a few days (Ramade 1978).

As is known, the control of rodents harmful to the economy is carried out in the USSR by poisoning with zinc phosphide or Gliftor. The bodies of the dead animals become toxic for predators, so that animals such as ermine, pole cat, Siberian ferret, common fox, water shrew, hedgehog also die. More frequently, rodents that have taken sublethal doses of the chemicals become easy prey for predators (which are known to serve the function in natural conditions of killing mainly ill or weakened animals). Actually, in the experiments of R.I. Dmitriev (personal communication), the tawny owl (*Strix aluco*) without mistakes first caught slightly poisoned rodents.

The disappearance of willow grouse in the Urals, and the abrupt decrease in the number of partridges, quails, ducks, geese and cranes are the results of poisoning these birds with fertilizers or their pecking at poison-treated grains.

From 2.3 to 5% of water fowl in the world had died by the beginning of the 1980's as a result of lead poisoning. In Wisconsin (USA) more Pallas's fishing eagles (*Haliaeetus leucocephalus*) had died from lead poisoning (entering their organism with water fowl eaten by them) than from poisoning with pesticides. In England about 63% of the swans (1500 specimens) that died and were examined between 1973 and 1980 had been poisoned by swallowing fishing sinkers. In Denmark there can be up to 400 lead shots per 7 m^2 of bed surfaces in shallow water. Lead shots have been found in the bodies of about 14% of dead mute swans (*Cygnus olor*) found here between 1975 and 1979.

Bioaccumulation (biomagnification) of pesticides, lead and other pollutants in the organisms of game birds is sometimes so high that they become dangerous for human consumption. Such a situation was noted in certain Western states of the USA in the autumn of 1981.

Sometimes an important factor in the occasional death of animals is poisoning as a so-called sanitization or medical measure (the control of rodents or insects which are carriers of infectious diseases, etc.). Thus, treating certain flood plain forests (Côte d'Ivoire) by fivefold spraying with Deltamethrine (12.5 g ha^{-1}) and one-time spraying with Permethrine (40 g ha^{-1}) to control the tsetse fly resulted in the disappearance of 20% of the freshwater fish species (11 out of 55), one shrimp species (*Cardins africana*) and an abrupt decrease in the number of many Hymenoptera and Diptera species; however, the number of tsetse flies showed no change.

More than 20 years ago, researchers noted and paid attention to the pollution of seas by plastic objects. Plastic particles have now been found in the stomachs of 42 sea bird species, of all turtles and in 5% of the Greenland seals in the White Sea. Up to 30% of turtle mortality is caused now by swallowing plastic objects (Wehle and Coleman 1983).

4.2.5.7 Mortality in Wartime or in Military Exercises

In the late 1970's, military operations in Uganda resulted in the loss of 14,000 to 46,500 large mammals in the National Park Ruvensori, amounting to about 30% of the respective species. Significant damage was caused to the fauna of the Semliki Reservation and the Cabalega Falls National Park. The ecological consequences of military operations in Vietnam are enormous. Here the American troops applied the herbicides 2,4-D and 2,5-T as defoliants over an area of more than 1.7 million ha. About 10% of the trees were killed after single treatment, and 85 to 100% were eliminated after four treatments. To recover these forests without artificial human intervention takes at least 100 years.

As a rule in nature, several factors resulting in the death of individual specimens, populations or species have a simultaneous effect. Thus their interaction can cause serious negative results even when each of them occurs at a minimal level.

4.3 Importance of the Minimal Number of the Population

The population can exist only when there are enough individual specimens. Determining the minimal number of individual specimens for a population to be maintained over a long period is now becoming one of the most important problems in the practical protection of many species, either held in captivity or living under natural conditions. Several fundamental studies in this direction have been performed by geneticists over the last few years (see, e.g. Frankel and Soule 1981; Soule and Wilcox 1980).

First of all, in solving the problem of determining the minimal population number, two different problems emerge the first problem of short range — to maintain the population for some time under controllable conditions without decreasing its viability and without (or with minimal) possible irreversible genetic or phenotypic variations. It is important to maintain the level of genetic variability, since the loss of such variability would prevent the solution of the second (long-range) problem — generating in the population the adaptation for living under the conditions of the changed environment.

In determining the minimal population number it should be borne in mind that not the total number of living specimens is important, but the number of propagating specimens, the so-called "the effective volume (number) of the population" (N_e). This volume generally equals 0 to 85% of the adult specimens in large (in number) populations and depends on the genetic contribution of the propagating specimens into the genetic pool of the subsequent population, variations of number, sex ratio, panmixia level. For example, if in nature in a population of 100 specimens there are 10 males to 90 females, then the effective population volume may be about 30 specimens. The effective population volume diminishes abruptly when the number fluctuates, which is inevitable in the life of every population; if the number of the population for ten generations ranges from 50 to 1000 specimens, then N_e may be about 300 specimens. Overlapping of the generations also induces a sharp decrease in N_e. These simulation estimates are only approximate, since they were made for the condition of random pair association during coupling, i.e. for the 100% panmixia (a condition that is never met in nature, and real computations are still impossible to perform, given the contemporary level of knowledge. The minimal population number is determined mainly by the danger of developing the negative consequences of the inbreeding, i.e. closely related hybridization. Classical population genetics and selection practice long ago showed three main dangers associated with inbreeding: (1) the appearance of so-called inbreeding depression; (2) the appearance of occasional changes in the phenotype; (3) lowering the hereditability of features and properties. The inbreeding depression manifests itself mainly in a decreasing growth rate, diminishing size and viability of adult species and loss of fertility.

The inbreeding ratio is equal to the rate of heterozygosis loss (or homozygosis increase), counted for one generation. It was shown that 10% increase of the inbreeding ratio can give rise to a 5 to 10% decrease in fertility. The 10% inbreeding ratio corresponds approximately to the inbreeding level that would be observed in a group of five propagating adult specimens for one generation, or of 25 propagating specimens for five non-overlapping generations.

Cattle breeders consider that the inbreeding level for one generation should not exceed 2 to 3%, otherwise the selection does not permit eliminating unfavourable alleles. The experience of stock breeding suggests the possibility of safe short-range inbreeding of 1% per generation (that corresponds to the effective population number of 50 specimens). This gives the rule for the short-range survival, i.e. "the rule of 1% inbreeding".

However, to maintain the wild animal phenotype it is necessary to select using a much greater number of features and properties than is customary in stock breeding. Increasing the number of features for selection makes it necessary to decrease the inbreeding level. After 20 to 30 generations, the population with N_e = 50 specimens will lose about one fourth of its total genetic variability.

A further practical conclusion is the marked decrease in fertility in small-sized populations even at inbreeding ratio values of about 0.5 to 0.6%. The empirical generalization is that the number of generations up to extinction as a result of inbreeding depression is about 1.5 times higher than the effective population volume (i.e. the population of ten adult animals propagating in each generation will be extinct after 15 generations due to inbreeding depression).

Unfortunately, the fate of the population is affected by other factors in addition to inbreeding depression and first of all to the above-mentioned fluctuation in number. In 1905, S.S. Chetverikov first showed the general nature and evolutionary importance of "waves of the life" i.e. the inevitable fluctuations in number in every population under the action of various ecological causes (e.g., see Chetverikov 1983). It is usual for the populations of the majority of higher vertebrates to change by 10- to 100-fold (larger number for small forms) and for the populations of invertebrates to change by 100- to 1000-fold.

As a result, it is assumed that the effective population volume for higher vertebrates that ensures their survival should be at least several hundreds of specimens, and for invertebrates, at least several tens of thousands of specimens.

As calculations and experiments in vertebrates and invertebrates have shown, the number of the present population is of less importance for survival than the number of specimens in the group to be maintained subsequently. The population of fruit flies established by a single pair and then maintained at this level for ten generations has retained the larger part (75%) of its genetic variability. The level of genetic variability in the population which was established by a single pair, but was maintained subsequently at the level of 10 to 20 pairs, showed no substantial differences in comparison with the level of variability for populations established by 10 or 50 pairs of specimens. The conclusion that the number in the maintained group is more important than the number of the establishing specimens is of extreme importance for the practical task of maintaining rare forms in captivity.

The problem of the minimal number of specimens in the population has not received due attention, as it is perhaps the most urgent problem in the protection of living nature as far as the population level is concerned.

4.4 Conclusions

Only some of the most important factors of negative anthropogenic action at the population and species level have been described in this chapter.

As a rule, in nature, several factors that result in decreasing the specimen number in populations and ruining the species as a whole act concurrently. In their interaction (synergism) they can give rise to serious negative results even at a low level of expression of each of them.

This situation is aggravated by the addition of the actions of various mutagenic agents (chemical exposure, irradiation, etc.) in the series of generations: the mutation burden in the populations under strong anthropogenic influence can apparently increase abruptly from one generation to another for many living beings.

5 Problems at the Ecosystem Level

To analyze conservation problems at the ecosystem level it is necessary to investigate inter alia problems of changes in their structure, disturbances of interspecies interactions, breakage of information flows, wreckage of vegetation cover in toto, transfer of pollutants by migrants and through trophic chains, and decrease of primary production.

5.1 Changes in Ecosystem Structure

Changes in species diversity and species composition seem to be among the most universal responses of ecosystems to any substantial disturbances.

The pollution of ecosystems, water and air induces numerous alterations in species diversity and composition.

In aquatic ecosystems, both chemical and thermal pollution may lead to extinction of species of fish, invertebrates, and phytoplankton (Fig. 5.1). Some taxons of animals are more sensitive to pollution than others. In some experiments species of Crustacea, Mollusca and Porifera were more sensitive to low concentrations of organochlorides than other taxons.

In terrestrial ecosystems pesticides are especially potent in inducing changes in species composition and diversity (Fig. 5.2). In one of the field studies it was found that treatment of forest (*Pinus* sp.) by the herbicide 2,4-D led to the ca. 50% reduction of the number of bird species and to ca. 50% reduction of the total number of birds.

Air pollution may cause very different changes in ecosystem structure. In Ohio (USA), it has been shown that the reduction of species diversity in trees of the upper stratum of the forest leads to an increase in light penetration and development of the strata of bush plants and herbaceous vegetation. That is only one example of a rather common situation when the exclusion of certain species from the ecosystem causes an increase in the abundance of others.

Not only pollution, but also other man-made factors induce alterations in the ecosystem structure. Among these are different physical (or mechanical) factors.

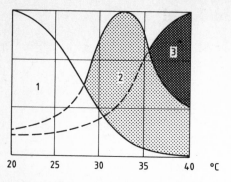

Fig. 5.1. Changes in the species structure of a reservoir's ecosystem caused by heat pollution. (Ramade 1978). *1* diatoms; *2* green; *3* blue-green

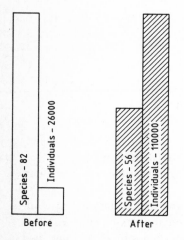

Fig. 5.2. Effects of chlororganic pesticides on the species composition and population of ground invertebrates in the meadow biocenosis. (After Ramade 1978)

Thus ploughing induces an increase in the number of some groups of soil mites (prostigmatids and astigmatids) and a sharp decrease of cryptostigmatid mites. Over-grazing and recreation usually degrade the terrestrial vegetation and impoverish the soil fauna of invertebrates.

The structure of plant communities in Europe and North America in the vicinity of human habitats and roads is changed due to synanthropization. Poland seems to be among the most carefully studied regions from the viewpoint of exploration of synanthropization of vegetation. It has been shown that only 3% of the country is covered by phytocoenoses with a low degree of synanthropization. In Poland, 33.5% of the territory has a vegetation cover totally changed or replaced by secondary agricultural, horticultural or ruderal vegetation.

5.2 Disturbance of Interspecies Interactions

If the number of species in an ecosystem is equal to N, then the number of possible connections between them may reach $N(N-1)/2$. For example, if an ecosystem possesses 1000 species, then the number of direct connections between them may reach 499,500. For comparison, a lake ecosystem may have more than 1000 species (for instance, Lake Onezhskoe has ca. 2000 species, Lake Vyrtsjarv has 1398 species; both lakes are located in the European part of the USSR). The Azov sea ecosystem is composed of more than 6100 species.

Several kinds of interspecific connection have been most extensively studied: the connections between prey and predators (including the links between phytophages and plants), symbiotic connections including parasitism, commensalism etc., and also the interactions between species which competitively use the same resource (feed, habitat, light) etc. The analysis of these interspecies relations is well represented in the literature. Comparatively less attention has been devoted to such consequences of anthropogenic effects on the ecosystem level as the transfer through trophic chains of pollutants, transfer of toxic substances by migrating animals and disturbance of ecological links due to the breakage of information flows.

Some conservation problems connected with interspecies interactions are considered below.

5.2.1 Disturbances of Trophic Chains and Other Interspecies Connections

A common situation in ecology is when a species B is crucially dependent upon a species A. In this case, one of the causes of decline and disappearance of the population of species B and finally of the species B itself may be the disappearance of species A. Examples of this are extremely numerous.

In Scandinavia the extermination of birds of prey induced the immediate stimulation of a number of game birds followed by a decrease in these birds due to an epidemic of coccidiosis caused by *Eimeria avium*.

Another example: the disappearance of fish in some African lakes after the killing of hippopotamuses. The excrement of hippopotamuses had been the natural nutrient for phytoplankton, which in turn had been the basis of the trophic chain: phytoplankton – zooplankton – fish.

In the 1950's on Bali (Indonesia), the major part of the dwellings was sprayed with DDT to exterminate mosquitoes. The latter disappeared, as well as the lizards that had eaten the mosquitoes. Then the cats which had eaten the poisoned lizards also disappeared. The dying off of the cats induced an enormous increase in the number of rats, and associated outbursts of plague. To save the situation, large numbers of new cats were shipped to the island, but too late, because some buildings had already been destroyed by the unusual activity of termites. The

increase in the numbers of termites had in turn been stimulated by the extermination of the lizards.

Man-made effects on ecosystems may lead to other disturbances of the
normal trophic and other connections between species.

Thus, the introduction of pesticides into ecosystems may pervert or change
the normal food preferences of phytophages and hence change the normal
pattern of trophic chains in the ecosystem. For example, observations have shown
that pesticide treatment made several normally neglected plants, including
Deschampsia caespitosa and *Senecio jacobaea*, more attractive for ungulates.

Tragic for many species and ecosystems are the disturbances in connections
between pollinators and the plant species that they normally pollinated. A
substantial part of the flowering plants are adapted for pollination by animals and
often by a narrow circle of pollinators (insects, birds, bats). Thus, in West
Australia alone more than 560 native plants (16% of flora) are pollinated by birds.

Reciprocally, many species of pollinators are specialized on feeding on
specific plants. The disappearance of the latter automatically leads to the end of
the existence of these animals. Thus, reciprocally essential for the survival of
populations of *Agave palmeri* and of their pollinators, the bat *Leptonycteris nivalis*
is simultaneously declining in the south of the USA and in the north of Mexico.
Simultaneous extinction of the animal and plant endemics — birds of the family
Drepaniidae and plants of the genus *Hibiscadelphus* — are occurring in Hawaii.
In both situations the plants had been the source of the food (nectar) for their
pollinators; both components of these couples are becoming extinct. Other
similar examples for the neotropical forest ecosystems of South America have
been discussed by Gilbert (1980).

The dependence of species B upon species A may be of a different type:
species A may provide for species B food, habitat, inevitable chemical "raw
materials" — ecological chemomediators etc. (Ostroumov 1986). The situations
are especially important when the dependence of B upon A is both obligate
(inevitable, vitally essential) and specific (i.e. species A cannot be replaced by any
other species). The examples of a sort of ecological cluster (Ostroumov 1986)
formed by such closely associated species A and B are very numerous (Ostroumov
1986). For example, all obligate monophages that are specialized for one species
of prey, all parasitic species that are narrow specialists and live on one single host
species, all narrow specialists among pollinators, all narrow specialists which
accept chemically inevitable molecules from one or a few species of plants
(Harborne 1982; Ostroumov 1986) etc., all these species associations represent
some sort of ecological cluster which is of importance for the assessment of
conservation problems.

5.2.2 Disturbance of the Equilibrium Between Species

To exterminate a fungus causing disease in an apple tree in one of the regions of
Canada, the plantations were treated with sulphur. The chemical treatment

inhibited not only the fungus but also the entomophages *Aphytis mutilaspidis* and *Hemisarcoptes malus*. The latter had been important natural regulators of a number of dangerous pests.

To protect the apple trees from pathogenic fungi carbamate pesticides were used which led to the increase of dangerous mite pests. Unexpected effects have sometimes been caused by these pesticides because they killed the natural enemies of the mite pests, the acariphages *Typhlodromus tiliae* and *Haplothrips faurei*.

Another example is the use of Hexachloran against locusts in Morocco. This led to a decrease in important enthomophages and one result was the increase in pest species of mites, *Coccodea*, *Aleurodoinea* etc. (Tischler 1965).

In 1981–1982 aerial Malathione spraying in California was conducted in an effort to eradicate the newly introduced Mediterranean fruit fly or Medfly (*Ceratitis capitata*). Among the negative ecological impacts of Malathione sprays were: increased whitefly parasite mortality and increased garden whitefly outbreaks, ice plant (*Carpobrotus* spp.) scale outbreaks due to increased mortality of ice-plant-scale parasitoids (*Encyrtus saliens*, *Coccophagus lycimnia* and *Metaphycus stramineus*); olive scale (*Parlatoria oleae*) outbreaks due to intoxication of olive scale parasitoids (*Aphytis paramaculicornis* and *Coccophagoides utilis*) (Dreistadt and Dahlsten 1986).

These examples indicate how a simple and frequently anthropogenic influence may disturb the natural equilibrium between species.

Anthropogenous reductions of a number of species and changes in species composition in ecosystems may lead to the disorganization of these ecosystems and to a decrease in their stability.

Classical evidence is the experiment with the predator sea star *Pisaster ochraceus* in the littoral ecosystem. The extermination of the species from the ecosystem led to a sharp reduction in number of species of invertebrates from 15 to 8. Probably this predator limited the number of one of the most effective competitor bivalves, *Mytilus californianus*, and hence created better conditions for other species that were in competition with *M. californianus*.

Other relevant facts were obtained during the study of sea mammals.

Overexploration induced a substantial decrease in numbers of several whales (*Physeter macrocephalus*, *Balaenoptera musculus*, *B. edeni*, *B. physalus*, *B. borealis*, *Megaptera novaeangliae*). This trophic link is now represented mainly by one species, *Balanoptera acutorostrata*. This is paralleled by the large fluctuations in number of krill which is food for the whales. Possibly the extended fluctuations are explained by the disturbance of the normal trophic web. It is possible that the disbalance of the normal trophic chain led to a decrease in the average level of krill in Antarctica and in turn to a further instability of whale populations that fail to recover in spite of their catching now being prohibited.

Another example is the bioenergetics of *Pagophoca groenlandica* in Canada. It has been calculated that both (1) increase in number of seals and (2) increase in catch of fish, which is the main food for the seals, will induce an unbalanced increase in energy cost of finding the fish by seals. A possible result may be the extinction of the seals because if they have not accumulated enough fat during the summer and autumn they may die during the winter.

Recently a huge amount of data about the delicate chemical regulation of equilibrium between species in native ecosystems has been accumulated by the application of biochemical approaches; chemical pollution may have additional detrimental effects through the disturbance of these natural regulatory mechanisms. Some further examples are given in the next section.

Anthropogenic disturbances of equilibria between species are possibly among the most common dangerous but often neglected and underestimated results of man-made effects on the ecosystems.

5.3 Breakage of Information Flows in Ecosystems

It is known that optical, chemical and acoustical signals are widely used by organisms as carriers for information flows in ecosystems. Hence ecological links may be destroyed by light, chemical and noise pollution.

Recent evidence is impressive that secondary metabolized and other natural compounds may serve as an intermediary (ecological chemomediators) between organisms or species; they may also serve as ecological chemoregulators of different processes in ecosystems. In order to better understand these signal links between organisms in aquatic ecosystems, it is interesting to consider new data on chemical attractants of the American eel (*Anguilla rostrata*) which detect odorants at concentrations of ca. 10^{-18} M.

Olfaction seems to be important to the migratory behaviour of many fish, including eels, but virtually nothing is known about the nature of the odorants involved. It has been found that stream bed gravel and stones, dead alewife eggs and river bank mud are attractive for migrating elvers. Sorensen (1983) showed that decaying leaf detritus (mostly from oak and beech trees) collected from the stream bed was attractive for elvers. Evidence was obtained that microorganisms either accelerate the release of the attractant or synthesize it themselves. We may conclude that cutting the riverside trees or pollution of the stream may break down the process of obtaining necessary information by fish.

In our experiments (Kaplan and Ostroumov 1986) we have obtained new evidence that chemical pollution may switch off the reception of chemical signals by fish. The anionic detergent Sulfonol dissolved in water was shown to inhibit normal electrophysiological responce of bulbus olphactorii of *Cyprinus carpio* to the chemical signals represented by L-alanin.

Another kind of danger is the disturbance of information flows between individuals of the same species owing to man-made effects on the habitat.

This has been shown in experiments with pheromonal communication in fish. The addition of phenol to water decreased the sensitivity of juveniles and adults of *Rhodeus sericeus* to their alarm pheromone, as has been demonstrated by Ponomarev et al. (1983).

In conclusion, it should be stressed that man-made interruptions or distortions of information streams in ecosystems may contribute considerably to the decrease in stability of ecosystems, changes in their composition, explosion-like growth of populations or extinction of a species. Effects of these types and genesis may be of great significance in the emergence of convenient and new pest species in agriculture on account of pesticide use.

5.4 Destruction of Vegetation Cover in Toto

Several kinds of human activity result in total destruction of native vegetation and hence native ecosystems cover.

The main causes are urbanization, mining and transport development. Millions of hectares of ecosystems are lost to concrete and asphalt annually.

The US are annually losing ca. 3 million acres of rural land owing to the spread of highways, housing developments and the like.

In Japan from 1968–1974, its annural rate of expansion in urbanized areas and industry was ca. 55,000 ha or 0.15% of the land area of Japan. According to the World Conservation Strategy, WCS (1980), from 1960–1970 Japan lost 7.3% of its agricultural land due to urban expansion. In Europe in 1960–1970 the loss varied from 1.5% (Norway) to 4.3% (Netherlands). In Iceland during Man's history, the total area occupied by vegetation cover dropped from 40,000 to 20,000 km^2.

Summarizing the situation in developed countries, at least 3000 km^2 of productive land with its ecosystems are annually submerged under urban sprawl (WCS 1980).

In developing countries, the similar trend is of a dangerous extent. Thus, in Egypt in 1955–1975 the loss of productive lands of the Nile valley and delta because of urban expansion was 400,000 ha.

The worldwide process of loss of productive land owing to non-agricultural use (i.e. urban and industry expansion) has not even been estimated with adequate accuracy.

Moreover, in the less developed countries (LDC's) an additional dangerous process exists: the rapid clearance and cutting of tropical rainforest at the rate 110,000 km^2 per year will destroy this type of ecosystem in toto within 80 years — and this means the demolition (and final extinction) of habitats of at least hundreds of thousand of species in the biosphere. Forest clearance often induces soil erosion and finally total destruction of vegetation cover.

It should be realized that any intense use of ecosystems or substantial transformation of them is equal to their destruction and loss of habitats for wildlife.

Native ecosystems or close-to-native lands are represented now only by archipelagos of relatively untouched areas; but even these remnants are not only under threat from further expansion of the human population and activity, of global pollution (see Sect. 5.7), of global climate change due to the greenhouse effect – but also they are regionally under threat of total physical destruction and annihilation.

More details are given in the section of this book devoted to forests, wetlands, islands etc.

5.5 Transfer of Toxic Substances by Migrating Animals

Migrating animals may transport different pollutants in their tissues; this problem has been only inadequately investigated and covered in conservation literature.

Thus, a significant number of birds fly back from Africa to Europe to their locations for reproduction; each of them may carry a small amount of persistent pesticides; as a result, the total sum of pollutants transported, e.g. of DDT transferred to countries where its use is banned, may reach substantial figures.

This point is of relevance to PCB, heavy metals, radionuclides and all other persistent pollutants, without exceptions.

This statement is in some degree connected with the famous example of V.I. Vernadsky (1967), who calculated the weight of only one of several flocks of *Shistocerca gregaria* in Northern Africa as over 40 million t. Such a flock may transport substantial amounts of Pb, Hg and other elements over distances of hundreds of kilometres; when eating vegetation sprayed by insecticides, a locust flock may carry and transfer to distant ecosystems a significant amount of accumulated pesticides.

Especially important at present may be the transportation by migrants of radionuclides which pollute the environment during emergencies and accidents in nuclear energy, or even during normal everyday functioning of nuclear stations and during the processing and storage of radioactive waters.

5.6 Bioaccumulation of Pollutants and Their Transfer Through Trophic Chains

To describe organism uptake and retention of pollutants, the following three terms can be most usefully applied. The definitions are given in the words of Connell and Miller (1984), with small modifications.

1. **Bioaccumulation** is the uptake and retention of pollutants from the environment by organisms via any mechanism or pathway.
2. **Bioconcentration** is the uptake and retention of pollutants directly from the water mass by organisms.

3. **Biomagnification** is the process whereby pollutants are passed from one trophic level to another, and exhibit increasing concentrations in organisms related to their trophic status.

The danger of bioaccumulation and biomagnification is underestimated.

Many authors include in their reviews tables with concrete figures about impressive bioaccumulation and biomagnification. However, on more rigorous study, it appears that most of these figures in their traditional presentations in books and even articles are given incorrectly: the recent data (e.g. Burdin 1985) demonstrated that the process of bioaccumulation is sensitive to a number of usually unrecorded factors. Their diversity includes: (1) water salinity, (2) food availability, (3) temperature, (4) speed of growth, (5) season and month, (6) depth of the organism's location, (7) concentration of other pollutants, (8) size and age of the organisms investigated, (9) length of time interval of exposure for pollutants studied.

Bioaccumulation of radionuclides is especially dangerous. Assuming a concentration of ^{32}P in water of the Columbia River (USA) equal to 1, it has been shown that the concentration of ^{32}P in phytoplankton (bioconcentration factor, BCF) is ca. 10^3, in fish usually consumed by Man up to 5000 (Ramade 1978). In phytoplankton BCF for ^{14}C was shown to be $> 10^3$, BCF of five other radionuclides (^{144}Ce, ^{55}Fe, ^{210}Pb, ^{31}P, ^{65}Zn) appeared to be $> 10^4$. Similar biomagnification has been demonstrated in terrestrial food chain: tundra soil → lichens *Cladonia* → *Rangifer tarandus* → Man. Owing to the bioaccumulation of ^{90}Sr and ^{137}Cs the dose of radiation accepted by local people consuming the meat of *Rangifer tarandus* was shown to be 55-fold higher than the dose of radiation received by inhabitants of Helsinki, according to calculations by Miettinen in 1965 (Ramade 1978). These results are in accord with those obtained in Alaska tundra ecosystems of the USA.

Numerous hydrophobic pollutants are being accumulated in tissues rich in lipids. During some periods, the reserves of fats and lipids are metabolized especially intensively. In the process of such effective catabolism of lipids, the pollutants previously stored in association with lipid structures may enter the blood circulation and poison other organs of the body. This is the case in some birds and bats which migrate over long distances. They may carry an amount of pesticides without any trouble in the begining of migrations; but at the end of their journey they sometimes die as a result of decrease of reserves of fat and a parallel increase in concentration of the pre-existing pollutants. Thus the level of DDT/DDE in the brain tissues of a Mexican bat, *Tadarida mexicana*, at the end of migration was shown to rise 40–160-fold.

The coefficient of bioaccumulation (bioaccumulation factor, BAF) is the ratio of two numbers, and it depends strongly upon the manner of determining these numbers. In many cases of metabolizable pollutants it is reasonable to take into consideration the sum of the concentrations of the pollutant itself and of its toxic metabolites (Fig. 5.3); this sum could then serve as a figure for calculating the ratio representing BAF. This is why the problems of bioaccumulation are

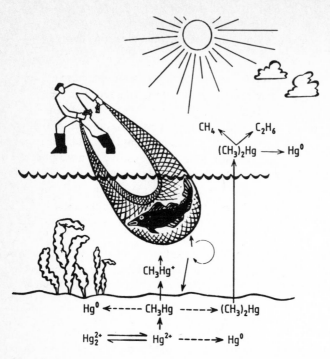

Fig. 5.3. Biotransformation and bioaccumulation of mercury in a water ecosystem. The *dotted lines* are reactions caused by bacteria (methylation of anthropogenic mercury in seabed sediments). Methylmercury accumulates in vertebrate tissues (the accumulation coefficient amounting to 500,000 as compared with the mercury content in water). The *round arrow* shows the passage of mercury along trophic lines. (Yablokov and Ostroumov 1985)

often inherently connected with the problems of biotransformation discussed above.

The transfer of pollutants in trophic chains is a dangerous process even without biomagnification.

The transfer of artificially produced radionuclides in the chain grass → cow → milk has been shown from a farm close to the nuclear fuel reprocessing plant at Sellafield, England (Sumerling et al. 1984). The transfer coefficient of transfer for activity from feed to milk (F_m) for ^{90}Sr was between 9×10^{-4} d l^{-1} and 4×10^{-3} d l^{-1}; the F_m was calculated as the ratio:

$$F_m = \frac{\text{concentration in milk, Bq l}^{-1}}{\text{daily intake, Bq d}^{-1}} .$$

These data and many other reports (e.g. Krivolutsky 1985; Kennedy et al. 1985) indicate the possibility of radionuclide contamination of the biosphere via trophic chains; the importance of this manner of transfer of pollutants increases due to the possibility of accidents.

5.7 Acid Precipitations

Wet deposition of air pollutants SO_2 and NO_x or their transformation products including solutions of H_2SO_4 and HNO_3 are referred to as acid rains (AR). Dry depositions may also play a negative role. The scale of acid precipitations (AP), as well as SO_2 and NO_x pollution, are discussed also in Chapter 1 and in Section 5.9 in this chapter. AP affect both terrestrial and aquatic ecosystems.

5.7.1 Effects on Terrestrial Ecosystems

In Europe, up to 7×10^6 ha of forest show signs of damage by AR (see Sect. 5.9). Besides forest die-off under strong pollution, another effect induced by low levels of AP is also alarming: the decrease in productivity. Tree growth (and associated productivity) may be decreased in association with annual mean concentrations 25–50 μg SO_2 m^{-3} which prevailed over large parts of Europe as long ago as 1982. According to some estimates, under pollution by AR, trees will not live more than 30–50 years (Hileman 1983).

Disturbances of soil chemistry and microbiology by AR are important. AP reduce bacterial numbers in acidified soils, diminish the diversity of microbiota. In some experiments acid treatments increased fungi and actinomycetes, e.g. an increase in number of *Phythium, Phytophthora, Penicillium spinulosum* has been detected. A reduction in the fungal mantle of spruce mycorrhizae under atmospheric pollutions was described Francis (1986). Acid rains may both accelerate and lower the rate of organic matter decomposition, and inhibit nitrification below pH 5.0. At pH below 6, denitrification is often inhibited and N_2O, instead of N_2, is released as the end product. AR inhibit asymbiotic bacterial nitrogen fixation in forest soils. Soil acidification also reduces symbiotic N-fixation. In some works the decrease of soil enzyme activity (cellulolytic, as well as urease and dehydrogenase activities) by soil acidification has been shown.

Probably AR may affect the biodegradation of some pesticides, e.g. Captan, Dicamba, Amitrole, 2,4-D, Vernolate, MCPA and several others were found to persist longer under acidic than under conditions.

It seems possible that AR associated with N deposition (in the Netherlands on average 15 kg N ha^{-1} per year) are the driving force inducing change from heathlands into grassland (Roelofs 1986). In particular, *Molinia caerulea* (L.) Moench and *Deschampsia flexuosa* (L.) Trin. expand at the expense of *Calluna vulgaris* and other heathland species.

5.7.2 Effects on Aquatic Ecosystems

AR affect species composition and diversity as well as energy flows through ecosystems, including primary and secondary productivities.

Acidification may alter the structure of the phytoplankton communities and reduce the number of species of Chlorophyta, Bacillariophyseae, and Cyanophyta. In acidified systems, diatom assemblages have become poorer in species; the relative abundance of *Eunotia exigua* may be increased particularly strongly. Acidification may cause new forms of dominance by dinoflagellates, diatoms and/or chrysophytes. In acidic lakes attached filamentous algae such as *Mougeotia, Binuclearia, Oscilatoria* etc. sometimes become abundant; this may inhibit the normal transport and cycling of nutrients in the ecosystems (Geelen and Leuven 1986).

Acidification of lakes and streams induces the disappearance of *Gammarus* and other amphipods, of larvae of the mayfly, *Baetis* spp. and the stonefly species, *Nemoura cinerea*. Among benthic organisms acidification reduces the populations of crayfish (*Astacus astacus*) in Scandinavia, *Orconectes viridis* in Canada, of the mollusks *Valvata macrostoma, Pisidium* spp., pearl mussel *Margarifera margarifera*, snail *Ancylus fluviatilis* and others (Okland and Okland 1986).

Acidic waters are unfavourable for vertebrates, also. Thus low pH raises the mortality rate of fish and of the eggs of the frog *Rana arvalis*. In acidic waters the egg masses of *R. arvalis, R. temporaria, R. esculenta, Bufo bufo* and *B. calamita* become heavily infested with fungi (Saprolegniaceae) (Leuven et al. 1986).

Acidification induces decrease in productivity of phytoplankton and zooplankton. It changes the normal energy flows through the main autotrophic components of the ecosystem – phytoplankton and macrophytes. During acidification, *Sphagnum* (usually terrestrial moss) has been shown to colonize the bottom of lakes. *Sphagnum* may reduce the number of macrophyte species *Isoetes lacustris* (*Pterydophytes*) which is able to fix CO_2 into malic acid in darkness (CAM metabolism) (Grahn 1986). Hence, acidification alters the normal pattern of CO_2 fixation and primary productivity.

The deposition threshold for damage to aquatic ecosystems was postulated to lie between 0.5 and 1.5 g S m^{-2} y^{-1}. The total sulphur deposition in some parts of the northeastern USA is at a level 1.6 g m^{-2} y^{-1} and even about double that in many regions.

5.8 Changes in Primary Productivity

At the ecosystem level substantial changes in the biosphere are connected with alterations in primary productivity (PP) induced by several anthropogenic factors. Different types of pollution are among the most powerful ones.

It is well established that many types of pollution decrease PP. In aquatic ecosystems this effect is exhibited by heavy metals, pesticides, PCB's, oils and petroleum products (Fig. 5.4). Our experiments have confirmed the ability of surfactants to inhibit the growth and hence the productivity of phytoplanktonic algae (Gorjunova and Ostroumov 1986). Inhibiting action on *Scenedesmus*

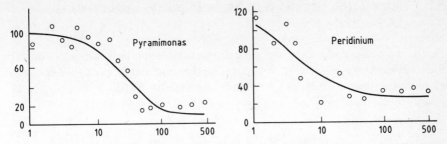

Fig. 5.4. DDT effects on the photosynthesis of seaweeds. *Y-axis* per cent of CO_2 absorption against the control figure; *X-axis* DDT concentration, bln⁻¹. (After Wurster 1968)

quadricauda has been demonstrated for anionic detergents sodium dodecylsulphate (SDS) and for sulfonol also (experiments by Kovaljova and Ostroumov, see Maximov et al. 1988a). Our experiments have shown that the surfactant Etonium induced bleaching of chlorophyll in the algae *Chlorella vulgaris* and *Monochrysis lutheri*, which is an indication of inhibition of PP.

Pollutants also inhibit photosynthesis, growth and PP of terrestrial plants. Air pollution by SO_2, NO_x, O_3 and others pollutants was shown to inhibit PP of many species of plants. Additionally, our experiments have found that surfactants may also inhibit the growth of plants and hence finally PP of phytocoenosis. Substantial concentrations of SDS, Sulfonol and Triton X-100 inhibited the development of seedlings of *Sinapis alba* and *Fagopyrum esculentum*; this meant that the oppressed plants were not able to be effective in PP (Gorjunova and Ostroumov 1986; Ostroumov et al. 1988; Maximov et al. 1988a).

The inhibition of growth of terrestrial plants induced by different pollutants that was mentioned in Chapter 3 reflects a concomitant decrease in PP. Thus, a recent review by Winner et al. (1985) has stressed the detrimental effects of SO_2 and NO_2 on growth, photosynthesis and productivity of *Picea abies*, *Pinus sylvestris*, *Fagus sylvatica*, *Tilia cordata*, *Betula pendula*, *B. pubescens*, *Populus nigra*, *Alnus incana*, *Malus domestica*, *Phaseolus vulgaris*, *Glycine max* and many others.

Both biospheric pollution and destruction of productive ecosystems (clearing of forests, drainage of wetlands, desertification etc.) seems without doubt to decrease the global PP.

On the other hand, the situation is not simple owing to the pollution of water bodies by N and P, which may induce eutrophication and local increase in PP.

In any case the result is that profound changes of patterning of PP over the Earth's surface are occurring and will certainly continue. Summarizing, their main causes are the following: (1) destruction of the most productive ecosystems, (2) conversion and shrinkage of vegetation cover, (3) pollution of air and water by substances inhibiting PP, (4) pollution of waters by N and P inducing eutrophication, (5) redistribution of water resources via irrigation and drainage, regional depletion of water supply, (6) possibly inevitable global and regional climatic shifts.

5.9 Conservation Problems in Some Important Ecosystems

To attain a better understanding of the dramatic deterioration of living nature at the ecosystem level it is necessary to consider the problems of biotic conservation in tundras, forests, drylands, grasslands, mountains, wetlands, ecosystems of rivers and lakes.

5.9.1 Tundra

Ecosystems of this type in several parts of world occupy ca. 8×10^6 km^2 or ca. 5.3% of the land surface. Both primary productivity and biotic diversity are low. In recent decades, the human population in the tundra has substantially increased.

Tundra ecosystems are being destroyed due to the exploitation of deposits of oil, gas and polymetallic ores. Tundra ecosystems are especially fragile, among the weakest elements being the vegetation cover. The moving tractor, off-road vehicle or cross-country vehicle may significantly destroy the vegetation cover and this destruction is either completely irreversable or the restoration of the vegetation may take tens of years. Over an area of 3 km the off-road vehicle may destroy ca. 1 ha of native ecosystems.

Ca. 65–67% of the total territory of the tundra is believed to be to some degree changed by antropogenic factors (Isakov et al. 1980).

Among the extinct tundra species is the bird endemic to the American tundra, *Numenius borealis*. Endangered are, e.g., *Grus leucogeranus* and *Branta ruficollis*. Under threat are also several other bird species, including: *Branta leucopsis*, *Cygnus bewickii*, *Eurunorhynchus pygmaeus* and *Rhodostethia rosea*.

Tundra ecosystems are insufficiently covered by reserves, zapovedniks[1], and other protected areas.

Among the most important problems of conservation of tundra biota are the following:

— the destruction of vegetation cover;
— overgrazing of tundra pastures by *Rangifer tarandus*;
— the sharp decrease in abundance (population number) of many animal species; ca. 20% of bird species are under anthropogenic depression of their number and ca. 5% are in an endangered state;
— the interruption of *Rangifer tarandus* migration pathways by oil and gas pipelines
— the accumulation of pollutants and waste material which are being biodegradated as an extremely low rate under the conditions of low temperatures and inactive microbiological processes.

---

[1] Zapovednik (Russ.) – a sort of reserve in the USSR with strict regime of conservation.

5.9.2 Forests

World forest land in 1981 was 31% of the world's land surface. Forest ecosystems are extremely rich in species and are very valuable both as their habitat and as one of most important life-support systems for mankind and the biosphere as a whole.

In 1950 the total forest area on the Earth was ca. 4×10^9 ha, in 1980 it decreased to 2.6×10^9 ha. If the present rate of deforestation continues, in the year 2020 the forested area will have shrunk to 1.8×10^9 ha. In 1978 in the world there were ca. 76 m^3 of wood per capita; in the year 2000 there will probably be only 40 m^3 per capita (Global 2000 1981).

Global deforestation is dangerously heavy and it influences even the global CO_2 level in the atmosphere. Due to forest clearing between 1860 and 1980 the global net release of carbon was between 135×10^9 t. At present the annual release of carbon due to deforestation is between 1.1×10^9 and 3×10^9 t. If deforestation rises in proportion to the human population, the release of carbon will reach 9×10^9 t/year before forests are exhausted in the next century (Woddwell et al. 1983).

A large area of forest land is expected to be reclaimed for agriculture because of the increasing demand for food. During a 25-year period, 1975–2000, the total area of forest land will possibly shrink by 15%, the forest land on potentially productive cropland will perhaps shrink by 55%; forest on highly productive land will probably shrink by even 70% (Buringh 1985).

An estimate of the present and future state of forest lands in different countries is given in Table 5.1

Among problems of forest ecosystem conservation, two different sets are considered separately: those of (1) boreal and temperate forest and (2) tropical forests.

5.9.2.1 Boreal and Temperate Forests

There are different types of classification of forests. Some authors distinguish, besides tropical forests, the following types: (1) cool coniferous forests, (2) temperate mixed forests, (3) warm temperate moist forests, (4) dry forests (which include also subtropical forests and partly bushes).

This section deals primarily with the first two types of these four, but the main conclusions may be extended to practically all types of boreal and temperate forests.

The species richness of forests is greater than that of tundra, and the biomass is also greater, reaching ca. 200–350 t ha^{-1} with annual not primary production ca. 10 t ha^{-1}.

Forests or the area of cleared forests are among the most densely inhabited areas of the world. Ca. 30–50% of this area is at present covered by anthropogenically transformed ecosystems and landscapes or merely urban areas. Up to 50% of this area in some regions of the USSR has been transformed into arable lands.

Table 5.1. World forest resources. (Global 2000)

	Closed forest (millions of ha)		Growing stock (billions m³ overbark)	
	1978	2000	1978	2000
Europe	140	150	15	13
USSR	785	775	79	77
North America	470	464	58	55
Japan, Australia, and New Zealand	69	68	4	4
Subtotal	1464	1457	156	149
Latin America	550	329	94	54
Africa	188	150	39	31
Asia and Pacific LDC's	361	181	38	19
Subtotal (LDC's)	1099	660	171	104
Total (world)	2563	2117	327	253

	Growing stock per capita, m³ biomass	
	1978	2000
Industrial countries	142	114
LDC's	57	21
Global	76	40

Among the present forest areas, the greater portion is covered by secondary forests or forest plantations which do not represent native indigenous ecosystems.

Physical destruction and transformation of forest ecosystems, chemical pollution and recreation are leading to the extinction of many species. In the FRG, a typically forest area located country, the numbers of extinct and endangered species are among mammals 8% and 47%, respectively, birds 8% and 38%, fish 6% and 70%, non-marine bivalve mollusks 3% and 32%, gastropod mollusks 1% and 21%, many orders of insects 3–13% and 5–60%, ferns and flowering plants 2% and 28% (plus more than 7% potentially endangered), mosses 2% and 10–14%, several groups of fungi 1% and 30–36%.

In the FRG 27 species of *Macrolepidoptera* (2%) are extinct and 494–534 (38–41%) are endangered. Extinct are 96 species of *Coleoptera*, and at least 1610–1686 are endangered (Blab et al. 1984).

A large part of Europe's woodland insects are now facing extinction en bloc. Of the 14,000 insect species of the British Red Data Book (see Speight 1985), 10% are either extinct or threatened with extinction. Of 1700 insect species surveyed in Belgium, 10% were either extinct or endangered and a further 25% could disappear by the year 2000. Insects of the fly family Thyreophoridae are almost certainly extinct throughout the whole of Europe. A minimum of ca. 10% of Europe's insects (ca. 6000 species) are threatened with extinction, and it is

reasonable to expect that by the year 2000 a further 12,000 insect species —
primarily woodland species — will be threatened with extinction in Europe
(Speight 1985).

Air pollution by SO_2, NO_x and possibly also O_3 lead to damage of forest
species, especially conifers. In Europe, damage appeared on silver fir *Abies alba*.
Norway spruce *Picea abies*, *Abies pectinata* and Scotch (common) pine *Pinus
sylvestris*, and other species also. In the USA, these pollutants induce damage to
ponderosa pine, *Pinus ponderosa*, easters white pine, *Pinus strobus*, American
white fir, *Abies concolor*, and also the angiosperm tulip tree *Liriodendron
tulipifera*, sweetgum *Luquidambar styraciflua*, and green ash *Fraxinus pennsyl-
vanica* (Ashmore et al. 1985).

In England, ambient air-induced foliar chlorosis has been shown on silver
birch, *Betula pendula*, European ash, *Fraxinus excelsior*, and two strains of white
ash, *Fraxinus americana*. The forest die-back (Table 5.2) owing to air pollution
and acid rains is so impressive that new expressions have been coined in
German: Neuartige Waldschäden ("new kind of forest damage") and Waldsterben (German term for forest die-off).

Effects of pollution on forests were studied in detail (e.g. Smith 1981; Ember
1982; Saith 1982; Shrinner 1982; Skelly 1982; Farthmann 1983).

Among forest ecosystems, riparian/alluvial forests which possess rich fauna
and flora are especially valuable and endangered.

In Spain, well-preserved riparian forests present 30–100% more species of
Papilionoidea and Hesperidea (and probably other insects) than adjacent areas.

Table 5.2. Development of damaged forest areas affected by acid rain in Europe. (Bach 1985; FAO
Forestry Department 1985; Bunyard 1986a and b and others)

Country	Year	Area damaged (ha or %)
ČSSR	1960	40,000
	1967	250,000–300,000
	1986	More than 30% of forested area
Poland	1961	80,000
	1968	240,000
	1973	260,000
	1980	379,000
GDR	1960	68,500
	1965	220,000
FRG	1907	9000
	1960	50,000
	1982	562,000
	1983	2,549,000 or 34.4% of forested area
	1984	3,698,000 or 50% of forested area
	1986	25% —"—
Italy	1985	5% of total
Luxembourg	1984	52% of forested area
Switzerland	1984	34% of forested area

The number of individuals has been shown also to be greater than 10–80% (Galiano et al. 1985).

Forests and other woodland ecosystems occupy 30–50% of the total land area of many countries (Table 5.1) and hence, the conservation problems of these ecosystems are extremely important.

Among the most serious problems of biotic conservation of boreal and temperate forest ecosystems are the following:

— collapse of the area of forests and other woodlands (and as result, destruction of habitats of many species);

— extinction and destruction of populations of plants and animals due to intensive agriculture and urbanization;

— contradiction between wood/timber production or harvest and the use of other functions of forest connected with water resource protection, air cleaning, recreation etc.;

— toxic effects of anthropogenic pollution including acid rains; modification and transformation of habitats (including old and dry trees, shrubs etc.) of organisms during logging, drainage, farming and sylvicultural practices, recreation and other forms of man's activities.

5.9.2.2 Tropical Forests

Tropical forests are an extremely important component of life-support systems (to use the WCS term). For Man, this type of biome is an important renewable resource. It has the role of a reservoir for genetic and ecological diversity; it provides a valuable supply of forest products if managed sustainably; tropical forests protect soils from erosion and are necessary for regeneration of soils. They protect regions down-stream from floods and siltation, contribute substantially to climate formation, global cycles of elements, biospheric CO_2 uptake and O_2 regeneration.

At present about 300,000 species of plants, vertebrates and invertebrates are found and described in tropical forests. The total amount of species in tropical forests is thought to be about 40–70% of the planet's stock of species, or somewhere between 2 and 5 million species altogether or even more (Wilson 1986). Probably no more than 15% of these species have been given scientific names, and most are completely unknown.

Tropical forests are the richest and oldest terrestrial ecosystems on Earth (the biota of some areas has a continuous evolutionary history since the end of the Cretaceous). This fact imposes an additional value on and explains many specific features of tropical forests.

There are a number of different classifications of tropical forest ecosystems which include about 15 types of these systems (e.g. Whitmore 1975; Poore 1978). Here we use the generalized expression "tropical forests" (as in WCS, Sect. 16), which sometimes means mainly tropical rain forests (Whitmore 1980) or tropical moist forests (Myers 1980a).

It is important to stress that the enormous biotic diversity of the tropical forest is often accompanied by another feature: low density of populations of individual species. For example, many species of plants are represented by very few individuals on a 10-ha sample. Hence, the populations are very vulnerable to extinction under man-made influence or as a result of fragmentation of habitats.

The total area of tropical moist forests has not been determined with certainty. According to Sommer (1976), the total primaeval climax area of tropical moist forest was 1600 million ha; the actual area in 1973 was about 935 million ha only. Other authors gave more uncertain estimates of 9–11 million km^2 (e.g. Myers 1980).

The unique role of tropical forests in the biosphere is determined by their huge biotic diversity and outstanding contribution to global photosynthetic CO_2 uptake and O_2 evolution and biospheric biomass. The average phytomass of tropical rain forests is in general about 400–500 t ha^{-1}. The net primary production is on the average approximately 30 t ha^{-1} y^{-1} (dry weight). Tropical forests fulfill 40% of global photosynthesis (e.g. Whittaker 1975).

The most important problem of tropical moist forests is the very rapid disappearance of this kind of biome and the reduction of its area. Sommer (1976) calculated that in Africa about 52% of the climax area of tropical moist forest has disappeared; in West Africa and East Africa even more, 72%. In Latin America 37% of the tropical moist forest, in Asia 42% and in South Asia 63.5% have disappeared. During historically recorded time a total of more than 40% of the climax area of tropical moist forest has been destroyed. As Poore (1978) concluded, the present area of tropical moist forests (evergreen, semi-deciduous) is still approximately 935 million ha or 58.5% of the climax area or 54.5% of the statistically reported total forest land area.

The area projected to be lost during the period 1975–2000 (in % of total closed forests 1975 area) is the following (WCS 1980): West Africa 47.1; East Africa and islands 17.8; centrally planned tropical Asia 29.1; South Asia 23.6; insular Southeast Asia 16.5; continental Southeast Asia 10.6; Central America 13.4; tropical South America 12.0.

The figures are even more dramatic if instead of total closed forests we consider so-called "operable" hardwood forests. "Operable" means 'productive" (for production of wood for industry) forests. "Inoperable" forests means a sum of the following types of forests: national parks, wildlife reserves, and other protected forests; forests on terrain that is too steep or wet; and forests which are without industrial wood potential. Areas projected to be lost in 1975–2000 (in % of "operable" hardwood forests) are (WCS 1980): West Africa 54.7; East Africa and islands 50.4; centrally planned tropical Asia 35.3; South Asia 27.9; insular Southeast Asia 26.3; continental Southeast Asia 13.3; Central America 23.9; tropical South America 13.3.

The main factors contributing to the disappearance and so-called conversion of tropical forests (which range from marginal modification to fundamental transformation of ecosystems) are definitive clearing of natural forest for colonization and agriculture, timber concessions and timber trade, forest farmers and shifting cultivation, cattle raising and firewood cutting. When all these factors were considered in conjunction, the rate of conversion of tropical moist forest was estimated to be approximately 200,000 km² per year. It is larger than the popular estimate of 105,000 km² per year or 20 ha disappearance per minute. The calculation made by Myers (1980a) is closer to the estimate of 40 ha per minute or 210,000 km² per year, offered by other authors.

Even if it is assumed that tropical rain forests are being felled and burned at the rate of 110,000 km² (Sommer 1976; WCS 1980), this will mean that at this rate all of this forest type will have disappeared within 85 years.

Myers (1983b) stressed that at present the conversion rate of tropical moist forests is higher — 245,000 km² per year (670 km² per day, 46 ha per minute).

Moreover, the lowland rain forests (the most valuable and richest in species) are being destroyed at a much faster rate. For example, the lowland forests of Southeast Asia are thought unlikely to survive the next 20 years (to the end of this century).

The average rate of tropical forest disappearance (in percentage) is believed by some experts to be between 0.7% and 1.2% of their tropical moist forest area (e.g. Poore 1978); for some areas the figures may vary from 1% to 5%.

Whitmore (1980) made his prediction of the expected degree of the tropical forest loss on the basis of expectations of tropical hardwood production for 1973-2000. Whitmore calculated that to meet the demand in hardwood at the 10 m³ per ha yield obtained in the 1960's and 1970's, consumption in 2000 will have to be 556 million ha of forest, or 59% of the 935 million ha remaining in 1973. As a result, this will leave only 30% of the total climax area under virgin forest. In this calculation no allowance has been made for definitive clearing of forest for colonization and agriculture and for fuelwood production.

The collapse of the tropical forest is the greatest source of large-scale danger for the Earth's biota. Some experts suggest that from 0.5 to 1 million species will be extinct within 20 years (by the end of this century). A major part of this possible loss is expected to be represented by species of tropical forests. Other dangerous consequences of the disappearance of the tropical forests are expected: the disbalance of the CO_2 cycle in the biosphere and the possible rise in its concentration in the atmosphere, changes of hydrological regimes in tropics, alterations of climate, not only of local character.

The actual total area of moist tropical forests in the world is about 10 million km² or even less. It is much less than the figure of 17 million km² which has been used in the widely accepted classical estimations of global net primary productivity (NPP). It seems reasonable to assume that the differences between these two figures (7 million km²) is an area occupied now mainly by agricultural lands, grasslands and secondary (successional) forests. Hence, the average NPP on this area (7 million km²) could be estimated at approximately 6–11 metric t ha⁻¹ year⁻¹ (Lieth and Wittaker 1975). On the basis of this it is possible to calculate that at present the NPP of the biosphere is being diminished by approximately 10^{10} tons per year, due to the decrease of this area of virgin tropical forests only. The total destruction of native tropical forests and their substitution by secondary forests, agricultural and grazing lands will more than duplicate this loss in global NPP. As mentioned before, at the present rate of destruction the total loss of tropical moist forest is expected within 85 years.

By 1979 in the tropical forest areas, 223 reserves and national parks were established. Their total area comprised 1.93% of the total climax area (or 2.99% of actual area of contemporary tropical forests).

At present among the most endangered and most valuable (as genetic resource) areas of tropical moist forests are the following: in Asia: northwest and eastern parts of the Malaysian peninsular; Borneo (especially exceptionally rich lowland, dipterocarp forests); Sumatra; the Philippines; Sri Lanka; Western Ghats in India. The lowland forests of Southeast Asia are expected to be unlikely to survive to the year 2000.

In Africa: Madagascar; so-called Pleistocene refugia of southwesters Ivory Coast; the relict forests of Ethiopia and East Africa's mountains.

In the Americas: western Amazon basin, the Pacific coast area of Columbia and Ecuador, and coastal and southeastern Brazil. In Brazil among others the softwood (coniferous) forests are projected to have a loss of 86% by the year 2000. Urgent conservation measures are required in softwood forests in Central America and in the Caribbean, since the pine species of these regions provide the genetic materials for the forestration programmes of many other countries.

Among 295 species of terrestrial vertebrates which are in the international Red Data Book (1978), 89 species inhabit the tropical forest.

There are many causes which make biota conservation in the tropical forest especially difficult, among them being: (1) The absence or weakness of comprehensive land use planning that takes adequate account of the need for the conservation of living resources within the context of socio-economic development, agricultural production and human settlement. (2) A lack of clear responsibility for exploitation and continuing management; uncertainty about the ownership of forest land. (3) Difficulties concerning the planned integration of indigenous peoples and shifting cultivators. (4) Institutional inadequacy and weakness in, e.g., legitimation, government structures and resources for education, training, enforcement and control.

It is necessary to stress that the real situation with tropical forests may be even worse than described here, because of lack of exact information. For example, according to official data, the forest area in India in 1980 was 74,743,000 ha (Baidya 1986), but satellite observation (1980–1982) gave smaller figures: total area under forest cover with a density 10% or more was 46,080,000 ha; with density 30% or more 36,020,000 ha (Rajan 1985).

In conclusion, the main problems of living resource conservation in the tropical forests include the following.

1. The rapid reduction (diminishing) of forest area. This leads to the extinction of a very large number of species and the possible disturbance of the global gas balance in the biosphere as well as to probable climatic changes, both local and global.

2. The necessity of rapid scientific, especially taxonomic, study of the species diversity in tropics. The scale of this challenging problem is unprecedented in the history of systematics, because it is required to describe many more species than are known in a number of taxons up to now.

3. The urgent need for the elaboration of an acceptable, ecologically reasonable, realistic and sustainable strategy of economic development in tropical areas and of usage of resources in this forest — a strategy which combines the interests of conserving the genetic diversity of tropical biota and the interests of development of the countries which possess these biospheric resources of international global importance.

5.9.3 Temperate Grasslands

There are a great variety of temperate grassland ecosystems: meadows, steppe (Europe, Asia), prairies (North America), Pampas (South America), high veld (South Africa), downland (Australia). In the USSR 14 types of steppe ecosystems, with different types of bioenergetics and of C and N cycling, have been described (Rodin and Bazilevich 1965). The eminent diversity of steppes and meadows includes 44 types of meadows, 24 types of steppes and six types of prairies.

In authoritative sources and textbooks the total area of meadow and steppe type ecosystems is usually estimated as 9 million km^2. Textbooks usually stress that steppe type ecosystems are unmatched in their floristic diversity, high level of humus content in soils (up to 7–8% and higher), fertility potential etc.

However, the classical grassland ecosystems of this type occupy this 9 million km^2 only on paper. Practically all these fertile ecosystems have been completely destroyed and converted to arable lands or transformed into pastures.

The value and rarity of the remnants of these ecosystems are so high that even the tiny remains with an area of 6–21 m^2 (and more) are recorded and are being investigated as sites of unique importance and scientific interest (Simberloff and Gotelli 1984); the largest prairie island among those remaining in Iowa and Minnesota (USA) was only 10^6 m^2.

A great many stenobiontic species of steppe ecosystems are endangered. In the USSR they include inter alia at least 16 species of *Stipa* (Poaceae), as well as the birds *Otis tarda, O. tetrax, Chlamydotis undulata, Burhinus oedicnemus, Chettusia gregaria* and many others (Takhtajan 1981).

The high speed of grassland ecosystem loss is illustrated by the figures calculated for Britain, which between 1950 and 1983 lost 95% of its lowland hay meadows and 80% chalk grasslands.

Among the most serious conservation problems of temperate grasslands are the following: (1) the highest degree of destruction in toto with loss of habitats and feeding for a great number of species, (2) rapidly increasing anthropogenic press on the remnants of steppe (in the broad meaning) ecosystems, (3) dramatic deficiency of protected areas.

Many important conservation problems of grasslands are common with those described in the sections of this book devoted to arid and semi-arid lands (drylands) and to agroecosystems.

5.9.4 Drylands Including Hyper-Arid, Arid and Semi-Arid Lands

Drylands occupy ca. 1/3 or 48×10^6 km^2 of the total land area of the world. (Treatment of drylands in this section is substantially based on the work by Dunford and Poore 1978). Of this, ca. 8×10^6 km^2 is hyper-arid, or climatic desert without vegetation except for ephemerals. The remainder is occupied by arid lands with sparse perennial and annual vegetation where nomadic pastoralism, but not rain-fed agriculture, is possible and semi-arid lands represented by steppe or tropical shrublands with a patchy herbaceous layer and perennial vegetation where rain-fed agriculture is possible. Among drylands ca. 2.5×10^6 km^2 are used for irrigated agriculture; 1.7×10^6 km^2 are used for rain-fed agriculture; ca. 36×10^6 km^2 are treated as pastures. More than 700×10^6 people live in dryland areas. Hence the anthropogenic stress on these ecosystems is huge.

The enormous diversity of dryland ecosystems is demonstrated by the following list: in North America the American desert (Colorado, Sonoran), the Chihuahuan desert; the Intermountain sage brush and creosote bush arid lands (*Artemisia tridentata, Larrea*); in South America Pacific coast deserts and semi-arid lands; dry savannas, shrub steppe, caatinga and semi-arid lands of Argentina, Bolivia, Brazil, Paraguay and Uruguay; in Australia arid lands of salt bush (*Atriplex, Maireana*), mallee, mulga (*Acacia aneura*), sclerophyllos grasses (mainly *Triodia*); South Africa the Kalahari semi-arid area; the karroo semi-arid area and the Namib fog desert; Africa and Asia the broad belt of drylands stretching across North Africa, the Arabian peninsula, the North West of the Indian subcontinent, West Asia and Central Asia; the belt incorporates: the Sudano-Sindian area including the Sahelian arid and semi-arid zone and savanna area fringing the Sahara to the south, and coasts of the Gulf and of the Red Sea, some areas in Ethiopia, Somalia, Kenya, India and Pakistan; the Saharo-Arabian drylands; the Irano-Turanian drylands including areas of the Turkish plateau, Syria, the Iranian plateau and of the Aralo-Caspian plain; the drylands of Kazakhstan and Dzungaria, with transitions to the European and Siberian steppe; the Central Asiatic drylands; the high altitude deserts of the Tibetian plateau.

The serious problem of these fragile ecosystems is desertification, a process accompanied by decrease in biological productivity with reduction of plant biomass and diversity and of the land's capacity to support livestock and crops.

Desertification is dangerous both for biota (destruction of habitats, extinction of species) and for the human population. Ca. 1/6 of the world population live in the drylands, and the actual number is growing quickly. Ten years ago more than 80 million people were threatened by a drop in productivity of drylands, of whom ca. 37% live in Asia, 30% in the Americas, 20% in Africa, and 13% in the Mediterranean basin. At present the actual number of threatened people is higher, probably more than 100 million.

According to some estimates prepared for the UN Conference on Desertification (UNCOD), desertification is proceeding at a rate of 60,000 km² (or twice the size of Belgium) per year. Globally, irrigated land is being degraded at a rate of 1250 km² year⁻¹, grazing land at a rate of 32,000 km² year⁻¹ and rain-fed cropland at a rate of 25,000 km² year⁻¹.

It has been estimated that 3.4×10^6 km², an area more than six times that of France, is under very high risk, 16.5×10^6 km² under high risk and ca. 18×10^6 km² under moderate risk.

The main causes of desertification are as follows:

1. Extension of irrigated agriculture into unsuitable areas or inadequate management of irrigated lands (Fig. 5.5). The rate of degradation of irrigated lands is comparable to that at which new irrigated areas are being brought into cultivation (in the developing countries the irrigated area is growing at the rate of 2.9% year⁻¹, dryland agriculture is increasing at 0.7% year⁻¹. Ca. 50% of dryland irrigated soils are suffering from salinization, 25% from erosion.

2. Extension of rain-fed agriculture into unsuitable areas. For example, in Nigeria before the recent drought, cultivation and settlements had been extended 160 km north of the recommended limit; in Tunisia during the period 1890–1975

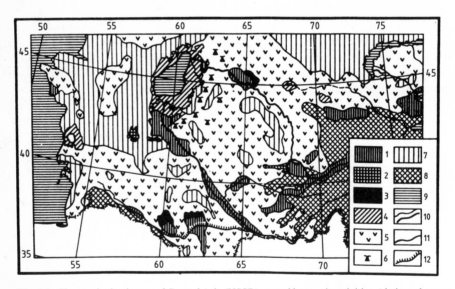

Fig. 5.5. Changes in the deserts of Central Asia (USSR) caused by man's activities. *1* irrigated oases; *2* discharge water reservoirs; *3* lands partially flooded and salinated by waters discharged from oases; *4* ecosystems suffering from aridization; *5* formation of tugai hydromorphic systems; *6–7* territories changing due to pasture irrigation; *8* non-irrigated territories; *9* mountains; *10* the Aral Sea coastline; *11* rivers; *12* canals. (Zaletaev et al. 1985)

27,000 km² of former steppe was cleared and brought under cultivation. In South Australia, a large area of salt bush with low rainfall has been ploughed for cereal cultivation. The result was degradation of the ecosystems.

3. Overgrazing and drought. The destructive combination of these two factors is the degrading of ca. 32,000 km² of arid and semi-arid land annually (about the area of Belgium).

4. Collection of firewood, which is a merciless threat for vegetation. In Gambia, collection for firewood occupies an average of 360 work days for the women per family annually. In the Sahel area the price of firewood is up to 25% of the family budget.

Many species of drylands are endangered. Several species are on the verge of extinction: e.g. Arabian oryx *Oryx leucoryx* and the ostrich *Struthio camelus* in Arabia. Many species have been diminished to a dangerously low level, among them Scimitar horned oryx, *Oryx dammah* and the Syrian wild ass, *Equus hemionus hemippus*. In Somalia there are at the lowest estimate 38 endemic succulent species of *Euphorbia* which are endangered. In Somalia and Ethiopia the valuable wild source of food, the Yeheb nut *Cordeauxia edulis* is endangered in the wild.

Among the species included in the USSR Red Data Book, at least 39% of mammals, 34% of birds, 73% of reptiles and many dozens of plants are inhabitants of arid and semi-arid lands.

The total portion of arid lands protected by national parks, reserves and similar protected areas has been estimated at ca. 1% which is completely insufficient.

Finally, the most serious problems of biotic conservation in drylands are: rapid desertification, reduction of biological productivity of primordially un-stable arid and semi-arid ecosystems, reduction of the carrying capacity of ecosystems and the destruction of fragile habitats; the destruction and exter-mination of populations of ungulates and many other species of animals and plants; extremely inadequate coverage by protected areas.

5.9.5 Mountain Ecosystems

Occupying about 25% of the world's land surface, i.e. ca. 40 million km², this group of ecosystems is of enormous diversity.

The superposition of zonality according to latitude and of gradients ac-cording to change of altitude produces a sophisticated mosaic of different habitats (often isolated) for animals, plants, fungi and prokaryotes. Mountain ecosystems are characterized by a high diversity of populations and species, and by a high amount of endemics. Mountainous environments served as centres for the evolutionary origin of many species, including those used in agriculture. The genetic diversity of montane species, including the wild relatives of modern crop cultivars, is of huge importance for work on the selection of new cultivars and hence for the prosperity of mankind. The mountain ecosystems are fragile and unstable; the habitats often have marginal conditions for many species.

The most serious conservation problems of these ecosystems include inter alia the following.

1. Deforestation. The decimation of forest ecosystems is a common phenomenon in mountains of all continents. For example, in the Hymalayas during the last 30 years, the forest area shrunk by 30%; in the mountains of Ethiopia 35 years ago 16% of the area was covered by forests, but in 1979 only 4% was represented by forest lands. The deforestation leads to a loss of habitats for many montane species and induces floods in the plains areas adjacent to the mountains.

2. Transformation of ecosystems of alpine meadows. These are totally destroyed or transformed to pastures with impoverished species composition owing to the pressure of livestock, which has markedly increased in number.

3. Destruction of ecosystems as a result of mining activities, construction of dams, reservoirs, roads etc.

4. Destruction of disturbance of ecosystems induced by recreational activities. The beauty, picturesqueness and wildness of mountains attract an increasing number of tourists for sport and all sorts of recreation.

The living nature of the mountains is in a more endangered state than many other types of biomes and areas. Thus, of the species included in the USSR RDB more than 30% species of mammals, about 20% species of birds, more than 80% species of amphibians and over 70% species of reptiles inhabit mountain areas (RDB of the USSR, 1984a).

In the mid-1980's in mountains the percentage of national parks and other protected areas was only 0.8% of total area. This is absolutely insufficient.

5.9.6 Rivers and Lakes

These ecosystems fulfill functions of the utmost significance: provide sources of water, of fish, of recreation and habitats for numerous hydrobionts which purify water by filtration. The ecosystems of streams are closely coupled with riparian ecosystems and this connection to a great extent determines the outstanding fertility and agricultural value of riparian lands.

The freshwater bodies occupy ca 0.4% of the surface area of the Earth and ca. 1% of the area of the continents.

All river and lake systems are insularized habitats for genetically isolated populations and hence the genetic pool of all river and lake inhabitants is unique.

The danger for hydrobionts is very great. In many regions fish species are among the most endangered taxons of vertebrates: thus, in the FRG, the total proportion of extinct and endangered species in fish is 71% (in mammals 53%, birds 52%, amphibians 58% and reptiles 75%) (Blab et al. 1984). Among the insects those taxons whose larval stages are located in river and brook ecosystems are the most endangered. According to Blab et al. (1984), the percent of extinct and endangered species is: Plecoptera 37%; Ephemeroptera 70%; Trichoptera 61%; Odonata 54%.

The extremely high level of anthropogenic pressure on river and lake ecosystems is determined and illustrated by the tremendous deficit of the sources of clean freshwater in the world: since 1975 ca. 60% of the population of developing countries has lacked adequate water supplies; ca. 4/5 of the world's rural population has lacked access to safe reliable drinking water; in urban centres ca. 1/4 of the people lack safe, reliable drinking water supplies.

The consumption of water by irrigated agriculture is huge, accounting for ca. 3/4 of total human water use, and it is expected to double worldwide by the year 2000.

Excessive water consumption leads to dramatic changes in some lakes, e.g. of the Aral sea, which is diminishing in size with the concomitant disappearance of habitats for hydrobionts.

Upstream manipulation of river systems, through dams or diversion for agriculture or municipal water supplies, destroys aquatic ecosystems and habitats as irrevocably as pollution, water transportation, recreation and modification of river banks.

Owing to dams, the natural reproduction of a great many species of fish has been inhibited, e.g. reproduction of several species of Acipenseridae (*A cipenser*, *Huso huso*), Salmonidae (*Salmo salar*, *Stenodus leucichthys*), and Coregonidae (*Coregonus lavaretus*).

Stream channel modifications have also played a role in the destruction of habitats of hydrobionts. Dredging of stream bottoms has proved harmful and deleterious for benthic communities. Streams were often redesigned, realigned and even relocated.

It has been estimated that by 1972 in the USA alone over 200,000 miles of stream channels had been modified and the habitats thus destroyed.

Pollution is another crucially detrimental factor for aquatic biota (e.g. Black et al. 1980; Gaigher and Hamman 1980; Dagani 1980; Barus et al. 1981; Black et al. 1982; Hannah et al. 1982; Werff and Pryut 1982; Jarvinen et al. 1983; Pawar et al. 1983; Begley et al. 1984; Babich, Stotzky 1985; Smith et al. 1988). Water samples from several water bodies in Europe and North America were shown to be mutagenic. In urbanized regions in summer the input of sewage waters to rivers may exceed the amount of water in these rivers. In addition to municipal, industrial and agricultural pollution, less known sorts of pollution exist: e.g. in the USA over 48,000 miles (76,800 km) of river were affected by the draining of acid waters from areas where mineral mining was carried out (Wells et al. 1983). Besides constant everyday pollution, accidents in industry and in waste-storage plants may be catastrophic for aquatic biota. Thus, in 1982 in the Lvov region the Dnestr river was polluted by 4.5 million m^3 of salt solution with a concentration of 250 g l^{-1}. In 1986, serious pollution of the Rhine owing to an accident was reported.

The pollution of water bodies countributes to the decline of many species of fish. In Russia among them inter alia (RDB RSFSR 1983): *Acipenser baeri baicalensis*, *Thymalus arcticus baicalensis* infrasubspecies *brevipinnis*, *Cottus gobio* etc. Pollution contributes to the decline of many invertebrates also. The list of them includes: *Margaritifera margaritifera*, *Dahurinaia middendorffi*, *D. dahurica*, *D. sujfunensis*, *Middendorfinaia* spp. and many others.

Water pollution induces the extinction of invertebrates of other taxons (especially of crustaceans and insects) and also of plant species.

It seems that the present level of exploitation of world freshwater fish resources has reached its maximum level and probably even more. The harvest of naturally produced freshwater fish was ca. 10×10^6 t in 1975 and has not increased over the period 1975–1980 (Global 2000). This is an indication that further increase is impossible and the present state of fish populations seems to be unsatisfactory.

Overfishing of many species of fish has led to their being endangered for instance many species of Acipenseriformes and Salmoniformes. Among endangered species are inter alia *Mylopharyngodon piceus* and *Siniperca chua-tsi* (Red Data Book of the RSFSR 1983).

Transformations of those parameters of aquatic environment that are vitally important for the survival of hydrobionts are dangerous and common phenomena especially in lakes, where increase in salinity, decrease of pH, and eutrophication may represent important examples.

Thus in the Aral sea, the increase of salinity was from 9.6‰–10.3‰(1960) up to 15.2% (1978). In some places salinity reached 17‰. This leads to the disappearance of economically important fish species. By the year 2010 the sea will possibly cease to exist in toto.

Acid rains induced a decrease in pH (acidification) in many lakes. Thus, in 1984 in the USA, ca. 3000 lakes and 23,000 miles of streams were seriously endangered, almost dead or dead.

In the beginning of the 1980's, a great number of lakes in Norway were practically devoid of fish and an additional large number suffered from reduced fish stocks (Fig. 5.6). In Sweden at least 18,000 lakes are acidified and incapable of supporting fish.

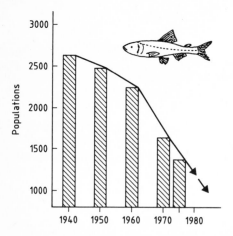

Fig. 5.6. Decrease in the number of brown trout (*Salmo trutta*) populations in South Norway from 1940 to 1980 as a result of the acidification of reservoirs. (Yablokov and Ostroumov 1985)

Acidification may induce the disappearance of fish via following mechanisms: (1) decreased food availability or quality, (2) recruitment failure, (3) fish death during episodic acidification.

Acidification induces numerous effects on aquatic ecosystems including reduction of species diversity of phyto- and zooplankton and even fungi; microbial litter decomposition is greatly inhibited in water affected by acid precipitation. Adverse effects of acidification were described in detail (e.g. Hendrey 1984; Bàker and Schofield 1985).

A crucially dangerous process is the eutrophication of lakes and other water bodies (e.g. Degobbis et al. 1979) which is initiated by pollution and increase in P and N; some species of algae produce enormous biomass and an excessive amount of organic matter; the quality of the water declines; species composition may substantially change.

A calamitous phenomenon for benthos is the pollution of sediments (e.g. Förstner 1984).

Another dangerous type of pollution is heat pollution by warm or hot water (e.g. Whitaker 1970).

In conclusion, the following problems are among the most serious in the conservation of living nature in rivers and lakes:

— increase of use of water for irrigation, municipal needs and for industry;
— construction of dams and all types of stream channel modifications and dredging of the stream bottoms;
— numerous sorts of water pollution;
— overexploitation of resources of hydrobionts including overfishing;
— intensive landuse and overgrazing of the riparian communities of rivers and streams;
— transformation of parameters of water bodies which are important for survival of hydrobionts; increase in salinity, decrease in pH, eutrophication.

5.9.7 Wetlands

These important ecosystems are of enormous diversity and it is not easy to define exactly which ecosystems are to be classified as wetlands. Some experts consider together "coastal and freshwater systems" including coastal wetlands and shallows, estuaries and coral reefs (WCS Sect. 2). Even with a more narrow approach there are many types of wetlands: (1) marshes, (2) swamps including mangroves, swamp forest, reed and papyrus swamp, (3) mires, peatlands (at least 500 million ha of the world's land), including landscapes of bog, moor, muskeg, and fen, (4) floodplains.

Coastal wetlands have been ranked as "life-support systems". Their most important and valuable functions include the following. They support food chains even beyond their boundaries due to currents and tides which export nutrients and food material. They maintain and improve natural water quality.

They provide habitats for a huge diversity of species both aquatic and terrestrial, many of which are either rare or endangered.

Thus, in the USA, about 35% of all rare or endangered animal species are dependent on wetlands or directly inhabit them.

Wetlands (including shallows) are essentially important for fisheries because they provide food, shelter, spawning and nursery grounds for fishes, crustaceans, and mollusks utilized by an estimated 2/3 of the world's fisheries. Moreover, in the Gulf of Mexico, 90% of commercial fish depend on estuaries and salt marshes.

Cash values of some wetlands are even higher than those of farmland. Thus, the total cash value of wetlands of the Charles River (USA) was estimated as US $65,600 ha^{-1} (including flood prevention, pollution reduction, water supply, recreation and local amenity); the value of salt marsh in southeast USA was estimated as US $212,500 ha^{-1}; those of coastal marsh in Georgia up to US $125,000 ha^{-1}. The average real-estate value of US farmland is 10,000 dollars ha^{-1}.

In Europe salt marshes provide habitats for about 1650 animal species with a high density of biomass and over 300 species of halophilic plants. Salt marshes protect valuable species of mammals and birds, e.g. in the French Camargue, Coto Doñana, Wadden Sea, the lagoon of Venice, to Norfolk Broads, heads of Scottish lochs and Scandinavian fjords. Salt marshes are very important in the press of biodegradation of pollutants: they treat sewage and are able to receive about 100 t ha^{-1} of silt and organic matter annually.

The total area of wetland ecosystems is shrinking dramatically. Thus, during the last five centuries, the UK's salt marshes have shrunk from 1000 to 450 km^2. The habitats of brant geese (*Branta bernicla*), widgeon (*Mareca penelope*), teal (*Anas crecca*), redshank (*Totanus totanus*), shelduck (*Tadorna tadorna*), common reed bunting (*Emberiza shoeniclus*) and others are being destroyed (Gissy 1986).

The destruction of wetland habitats may lead to a decrease in the populations of migrating birds that use these wetlands during their migrations. Thus, South America wetlands had been used by the migrating bird *Numenius borealis*; the agricultural development and concomitant destruction of these wetlands have led to the extinction of this species.

Among various sorts of human activities in wetlands is peat mining, which may lead to serious environmental problems (e.g. Winkler and DeWitt 1985):

1. When peatland soils are drained the adsorbed toxic metals (Hg, Pb, Cd, As, Zn) are released and carried in ditches to rivers, lakes and coastal estuaries.
2. Drained peatland runoff in the USA (North Carolina) had three times the N and 28 times the P that was found in runoff from undrained peatland. The result may be eutrophication of surface waters.
3. Peatlands are groundwater recharge areas and removal of the peat may contribute to reduction of groundwater.
4. When trees and peat are removed from peatlands, runoff increases and results are increased flooding and increased freshwater discharge into coastal

estuaries. Effluents with high heavy-water and acid content may have impact on hydrobionts and fisheries.

Among the most important conservation problems associated with wetlands are the following.

1. Total destruction of wetlands as ecosystems and habitats for many species. Within the last 100–200 years as much as 50% of the wetlands in the world has been lost (Winkler and DeWitt 1985). In some regions of the USSR (e.g. the Ukraine) all remaining wetlands will probably be lost within the next 30–35 years (Boch and Mazing 1979).
2. Change of species composition of wetland ecosystems, synanthropization of flora.
3. Eutrophication of wetlands due to discharge of runoff and of polluted waters. Results are changes of species composition.
4. Fires which often occur in the areas where the peatland has been drained.
5. Carbon losses from drained wetlands contribute substantially to global changes of carbon cycling in biosphere. Thus, carbon losses from only two peatland areas in the USA, in Florida and California, equal on an annual basis 0.3% of world CO_2 release from burning oil and coal.

Worldwide, the total protected area of wetlands is too small. According to the Ramsar convention, 192 wetland reserves exist in 29 states of the world; this is only 0.22% of the total area of world's wetlands. In the USSR, 2% of wetlands are protected, but this is insufficient. It is necessary to increase this figure to 10–15%.

5.10 Conclusions

We have only been able to touch upon several types of biomes and ecosystems. All of them show that rapid antropogenic transformation is everywhere present. It is also true that the numbers, general area, and geographical distribution of the protected areas is far from being extensive enough (Table 5.3).

Table 5.3. Analysis of the natural protection areas by biome type. (1985 United Nation List of Natural Parks and Protected Areas, Gland, Cambridge, 1985)

Biome	Number of areas	Total area (ha)
Tropical humid forest	280	39,052,051
Subtropical/temperate rainforests/woodlands	275	18,530,260
Temperate needle-leaf forests/woodlands	175	38,785,369
Tropical dry forests/woodlands	581	65,529,677
Temperate bread-leaf forests	483	11,526,977
Evergreen sclerophyllous forests	475	12,004,751
Warm deserts/semi-deserts	161	41,603,555
Cold-winter deserts	76	13,547,995
Tundra	40	115,467,889
Tropical grasslands/savannas	30	9,052,796
Temperate grasslands	90	1,933,838
Mountain systems	436	32,534,801
Island systems	270	15,819,748
Lake systems	10	518,578
Biogeographical classification unknown	132	7,867,578
World total	3514	423,774,398 (= 2.8% land area)

6 Economic Problems in the Protection of Living Nature

It has been repeatedly emphasized that the economic evaluation of living natural resources is not only desirable and important but also hard to effect. A few examples may elucidate some of the problems involved.

Much gold has been mined since the famous gold-rush in the 19th century in Alaska. Yet the total value of furs, fish and timber procured there in the course of that century may well have exceeded the value of the gold. Moreover, Alaska's surface gold deposits have been exhausted, whereas its biological resources can still bring in considerable profits. History repeats itself, now involving oil, large deposits of which have been discovered in Alaska and adjacent territory. But however large they might be, they, too, will be depleted in due time. The animal and vegetable kingdoms, however, will continue to yield profit indefinitely, provided, of course, they are preserved. Given these facts, how can animate and inanimate natural resources be compared? If a one-time comparison is made, the balance is usually tipped by resources whose value is higher at that particular moment, the future being disregarded.

The period for evaluating economic results may be determined in the following way. Four generations of people live simultaneously on the Earth. By giving life to the succeeding generations, the older one is committed, as it were, to pass on at least the same quality of life, including, of course, the quality of the natural environment.

If we take the life-span of one generation as equal to 25 years, the minimal estimated period will be 100 years. But a century is an insignificant period for evolutionary and biospheric processes. So, it, too, seems to be insignificant.

Another example shows how hard it is to make a true assessment of living natural resources. In the late 1970's the cost of timber from one hectare of coniferous forest in the USSR amounted to some 500 roubles. Calculations show, however, that the Sunday recreation of workers in suburb forests alone results in a 0.3% increase in the annual average productivity of labour at heavy engineering plants, or 91.7 roubles per hectare of the green zone around the town of Voronezh (USSR), where the estimates were made. The hydrology regulating the role of one hectare of forest in the central part of the USSR is estimated at 685 roubles. In the USA, the water drainage capacity of one hectare of forest is evaluated at US $ 1000 per year, while in Czechoslovakia the figure is K 10,900 per year. Forest recreational assessments vary widely, depending on locality and calculation method. Yet they may exceed by 20 times the cost of felled timber. The health-improving effect of each hectare of green plantations in Moscow and its urbanized surroundings alone is estimated at 4000 to 5000 roubles a year. The aggregate "social value" of forest, including its water regulation, purification and recreation functions, amounted to DM 4880 per hectare in the FRG, many times higher than the price of timber. The figure for the Moscow region is 6000–8070 roubles (Reymers and Shtilmark 1978; Fedorenko et al. 1980).

Yet another example refers to the exploitation of Africa's wildlife resources. It has been estimated (Conway 1980) that the cost of one lion is US $ 515,000 if kept on show in Amboseli National Park (Kenya), $ 8500 if used on a hunting farm and $ 1150 if killed and skinned. The cost of a domesticated lion used for 15 years equals that of 30,000 head of cattle.

From the socio-economic point of view, there are three groups of losses brought about by the destruction of the natural environment: (1) calculable economic losses; (2) incalculable social losses; and (3) lost opportunities involved in the future use of resources.

Examples of economic damage caused by air pollution have become common knowledge. Specifically, a twofold increase in air pollution reduces the durability of industrial equipment by 50%. Wheat capacity in zones affected by non-ferrous metallurgic facilities proves to be from 40 to 60% lower and the grain protein content drops by 25–35%. The number of lost working days and the cost of medical treatment form a basis to determine the extent to which people's health is affected by air and water pollution. According to OECD estimates, losses caused by air pollution as a factor contributing to the deterioration of people's health in the industrialized capitalist countries account for 3–5% of the GNP, amounting to dozens of billions of dollars. Unfortunately, there are no similar estimates with regard to wildlife.

It is most difficult to assess economic damage caused by lost opportunities involved in the future use of this or that resource. A pertinent example is the extermination of Steller's sea cow (*Rhetina stelleri*). The short-term benefits of its procurement cannot be compared with the lost possibilities of breeding this animal, which had delicious meat and could use the colossal resources of the seabed coastal vegetation of temperate seas. It is even more difficult to evaluate economically the loss of this species' genetic pool (20% of the genetic diversity of the whole order of mammals).

Another group of unresolved problems arises from the fact that a purely economic approach to the evaluation of a particular living natural resource is inadequate. A wild duck (*Anas platyrhynchus*) shot in autumn by a hunter in the USSR sells at 2–5 roubles. The same wild duck which has hatched ducklings in a pool right in the middle of a busy downtown square and which trustingly takes food from people, is important for fostering good will and gives aesthetic pleasure to the townsfolk. Its very presence makes the urbanized territory more adequate for man's biosocial needs, and so the value of the duck increases many times over. It is possible to calculate fairly accurately the cost of protecting ducks in town, feeding them and cleaning the territory, but that sum will represent only a tiny fraction of the bird's social value. Similarly, the evaluation of a thrush, a nightingale, a waxwing or a squirrel in a city park must be many times higher than their value in nature.

So, the traditional economic approach alone is not sufficient to give a true assessment of a living natural resource. In 1975, for instance, the extraction of minerals from the seabed and the ocean floor was worth nearly US $ 70 billion, world sea navigation brought in some $ 40 billion, whereas the value of marine products was estimated at as little as $ 14 billion. But does this mean that the extraction of marine minerals was five times as economically important as the provision of marine food protein for hundreds of millions of people throughout the world?

The discrepancy between the economic effort and the world community's real needs becomes all the more evident when the problem of arms expenditure is considered. In some cases such expenditure is more than 10% of an industrialized country's national budget. The military budgets of all major countries, in both absolute and relative terms, are several times more than their government outlay for the protection of the environment. More than US 600 billion is spent on arms annually across the world, or over $ 2 million every minute. This excessive military expenditure seems to be the prime reason for the shortage of finance to work out and implement large-scale programmes to protect and improve the quality of the natural environment and save animal and plant species from extinction. All this leads clearly to one conclusion: there is no reasonable alternative to detente, to the development and consolidation of the principles of peaceful coexistence, to a halt in the arms race and to disarmament.

Given the current rate of deterioration in the quality of the environment, only special state-sponsored efforts will make it possible to slow down the process somewhat. Allocation of at least 2% of the GNP for protecting the environment (meaning, first of all, air and water) will make it possible to reduce deterioration processes substantially in these components of the environment. If 6% of the GNP is appropriated for the protection of nature, deterioration processes will cease altogether, whereas from 8 to 10% will be enough to restore the purity of the atmosphere and water.

In the industrialized capitalist countries, national expenses on environmental protection measures vary from 1 to 2% of the GNP. In the USSR, such expenses (including those covered by industries and local authorities) amount to nearly eight billion roubles annually, of which 75% is spent to protect water, 12% air, 8% land, 5% the Earth's mineral resources, forest and animals (Fig. 6.1).

There is no acceptable method so far to make an economic evaluation of the damage by man caused to living nature as a result of violations of the existing rules and standards of nature utilization. Neither is there a method to make an adequate economic assessment of measures to maintain the ecological balance, including those to preserve the genetic diversity, landscapes, reserves and other protected natural territories.

The protection of living natural resources is in keeping with the basic and long-term interests of developing human society, and it cannot be uneconomical for society if only for this reason. If, however, it sometimes appears so, this is due

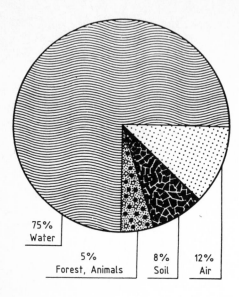

75%
Water

| 5% | 8% | 12% |
| Forest, Animals | Soil | Air |

Fig. 6.1 Distribution of expenditures for nature protection in the USSR in 1976–1980. (After Khachaturov 1980)

to the imperfect methods of economic evaluation used today. Can anyone speak of a high price when the preservation of life on this planet is involved?

To sum up, several general conclusions can be made regarding the economics of nature protection.

1. When evaluating inanimate resources and the resources of living nature, their basic difference is not fully taken into account. The value of the former is determined on the assumption of their one-time use, whereas the latter's value largely depends on their exploitation possibilities: their cumulative value in the case of inexhaustible protracted exploitation is many times higher than that in the case of their one-time, "one-salvo" use, similarly to inanimate natural resources.

2. A narrow, purely economic loss-and-profit assessment of living nature proves inadequate because it ignores the social and ecological aspects. The socio-ecological evaluation of animals and plants, and also individual ecosystems may differ radically from their narrow economic value.

3. The current methods of determining damage done to living nature and the efficacy of expenditure to protect the animal and vegetable kingdoms cannot be effective levers to improve the protection of living nature.

All these facts show clearly that economics must be ecologized.

7 Organizational and Legal Problems of the Protection of Living Nature

The basic organizational and legal problems involved in the protection of living nature can be divided into two groups: (1) those which can only be solved on the international level, and (2) those which can be solved on the level of individual states. The second group can be subdivided into problems confronting capitalist, socialist, and developing countries.

7.1 Problems to Be Treated on an International Level

It is common knowledge that the problems of environmental protection (the rational utilization of nature) are global issues confronting the whole of mankind. They can only be tackled by collective efforts made by the world community at large. Many aspects of the global protection of the environment pertain directly to living nature.

These problems can be divided into three groups. The first is connected with the utilization of the living natural resources of the World Ocean or, more broadly, the living natural resources beyond national jurisdiction. The second group deals with the exploitation and protection of migrating animals which spend only part of their life in habitats under the jurisdiction of one state (Fig. 7.1). The third group deals with maintaining the biosphere's overall parameters vital for the very existence of life on our planet (the quality of the atmosphere, the hydrosphere, etc.). The latter group has partially been considered earlier. The discussion of its organizational and legal aspects naturally goes far beyond the scope of this volume.

7.1.1 Living Natural Resources Beyond National Jurisdiction

These resources referred to as "natural objects beyond national jurisdiction," "international natural objects," "common resources of all humanity" or "shared natural resources" (WCS 1980) include the animal and vegetable worlds of the

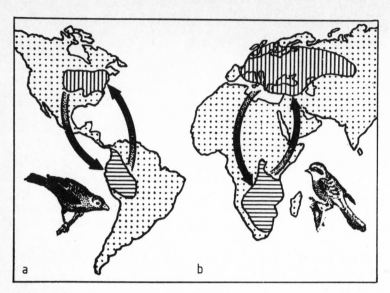

Fig. 7.1a,b. Many bird, cheiroptera, fish and invertebrate species cross national borders. The nesting habitats, migration routes and hibernation sites of: (**a**) scarlet tanager (*Piranga olivacea*); (**b**) brown shrike (*Lanius cristatus*). (Yablokov and Ostroumov 1983)

pelagial zones of the world's oceans and the Antarctic, the only continent not divided between states. As regards the ocean's vegetation, environmental problems involve mainly the protection of water from growing pollution. International measures to protect animals "beyond national jurisdiction" have long been urgent. Such measures involve primarily marketable fish, cetaceans, pinnipeds, and also invertebrates, such as cephalopods and pelagic crustacea.

In practice, there are three types of international rules for utilizing such living natural resources: (1) free access to natural resources; (2) their joint control; and (3) their international ownership (Bilder 1980).

The principle of free access to resources beyond national jurisdiction is realized in the form of freedom of fishing in the high seas. Unfortunately, this principle proves to be imperfect from the viewpoint of the protection of living nature. People obtain a considerable part of the animal protein that they need from the World Ocean, ranging from 10 to 20%. In recent years, the World Ocean's ecosystems have constantly been losing more than 70 million tons of marine objects, of which over one third is used to make fodder, technical and food fat. Modern fishing technology allows one trawler to reap a 1-year yield from 5 to 10 km² during 1 hour of trawling, while one purse net sweep can gather a year's harvest from several hundred square kilometers of water surface.

Intensive fishing in the World Ocean has undermined the resources of a number of species, once prolific. Although it has not brought individual species to the brink of extinction, intensive fishing on the principle of "freedom of fishing in the high seas" has affected individual marine ecosystems to a no lesser degree than intensive agriculture in the steppe zone.

The second principle of the utilization of living natural resources beyond national jurisdiction, i.e., their joint control by a limited number of interested states, has been effected regarding cetaceans, pinnipeds, and Antarctic krill (*Euphausia superba*). Let us consider this principle in more detail, taking the exploitation of big cetaceans as an example.

Some cetacean species are now on the brink of extinction. This applies primarily to the largest whales: Bow head (*Balaena mysticetus*), Right whale (*Eubalaena glacialis*) and blue whale (*Balaenoptera musculus*). In the 17th–18th centuries, Bow head and Right whale hunting, and also cachalot hunting, had long been a source of technical fat, replacing functionally modern oil production. These species were exterminated on a mass scale.

Taking into account the position of animals in food chains, the killing of one whale amounts to harvesting from 300 to 500 km² of ocean area. The extermination of whales seems to affect the entire ecosystem of the World Ocean even more than fishing. Being at the top of the ocean's trophic chains, whales and other sea mammals play an immense role in stabilizing this ecosystem, maintaining its stable equilibrium. It is quite possible that the drastic changes in the number of many species of marketable fish observed over recent decades are to a considerable degree due to the disappearance of whales, which fact reflects the on-going destabilization process in the entire ecosystem of the World Ocean. Sharp increases and decreases in the number of species are in fact characteristic indicators of destabilized ecosystems (Odum 1971). Yet it is impossible to predict whether the ecosystem at large would be able to stabilize at some other level, because contemporary methods of ecological prognostication are in principle impotent when it comes to forecasting the behaviour of complex ecosystems with many independent variables.

Experience shows that all of the World Ocean's living natural resources subjected to intensive exploitation began to shrink sooner or later. This impelled the states participating in such exploitation to conclude special agreements on the protection and rational utilization of resources. As of now, there are more than 80 international agreements pertaining to the rational utilization of the natural resources of the World Ocean. It might seem that, theoretically, the problem is solved: resources beyond national jurisdiction should be exploited with a longer perspective in view, without undermining the main reserves and living, as it were, on an "interest on principal." Unfortunately, the application of such agreements shows their inadequacy.

The International Whaling Commission (IWC), established within the framework of the relevant Convention (1946), has done a great deal to determine the economic load on various species in particular regions. Yet the IWC's activities have so far resulted only in delaying a ban on the hunting of virtually all regulated species (for 10–20 years, according to various estimates) and the saving of whales as a zoological species. Today, whaling has been stopped almost everywhere, essentially because it is uneconomic to hunt species which are extremely rare. We believe that the following factors are behind this deplorable fact:

1. the striving of some countries to reap immediate maximal possible profits, even at the expense of future benefits ("better less, but now");

2. the constant shortage of reliable scientific data to calculate adequately the maximum allowable hunting load on both particular species and individual populations;

3. the absence of adequate control over the execution of decisions (international inspection of national whaling facilities has been introduced only recently);

4. the non-participation of some whaling countries in the Convention and their non-compliance with IWC decisions, and also the spread of poaching;

5. the inadequate participation in the IWC's activities of non-whaling countries interested in the preservation of whales as an important component of the World Ocean's living nature.

The latter factor was somewhat corrected in recent years when some "neutral" countries joined the Convention, thus increasing the number of its participants. This has enabled the IWC to adopt important decisions to protect the remaining world whale resources. This development is extremely important: as long as the management of such resources was concentrated in the hands of countries having a direct interest in their exploitation, they could not agree on a long-term resource-protection policy, but sought to "grab" as much as they could, putting their immediate interest above that of the world community.

The inadequacy of the principles of "joint control" and "free access to resources" underlies the desire to declare the World Ocean's resources an international property, the property of mankind. The development of this principle seems to be most promising.

National attempts were also made to solve the problem of natural resources beyond national jurisdiction. In 1974, most littoral States announced a 200-mile fishing and economic zone. As a result, the area of the ocean beyond national jurisdiction has shrunk by approximately 130 million km², which amounts to 36% of its water area. These are the highest-yielding fishing zones, accounting for over 70% of the world catch of seafood. This "appropriation" or the establishment of national jurisdiction over a considerable part of the ocean may facilitate the organization of the long-term rational exploitation of marine resources. This, however, does not resolve all the problems of resources beyond national jurisdiction, which continue to be a potential source of danger for the preservation of many species of game animals.

7.1.2 Protection of Animals Migrating over National Borders

Another group of international problems is connected with the protection of migrating species spending only part of their life cycle on the territory of a particular country. One may organize perfect protection of any bird species at their reproduction sites (tundra, for example), yet the effect will be negligible if the species are actively exterminated at their hibernation sites. Relevant examples are plentiful.

Migrations are characteristic not only of birds, but also of some fish species (from sea to river and back), some cetaceans, pinnipeds and hoofed animals. Certain invertebrates, including some large day-time butterflies, also migrate. The presence of such migrating species, many of which are included in the categories rare, decreasing or under the threat of extinction, makes it imperative to regulate the nature conservation activities of those states which prove to be responsible for the future of such species. This is a typical organizational problem of an international nature, which for the time being is only in rare cases resolved adequately.

A general solution to this problem requires coordinated nature conservation efforts on the part of neighbouring states within large geographical regions.

7.2 Problems to Be Treated on a National Level

The protection of living nature on a national level depends on the socio-political system and the level of economic and social development of a particular country. The desire to preserve the rich variety of the vegetable and animal kingdoms is common for all countries. In the context of intensive economic development, this can only be achieved by organizing large protected natural territories with a special regime, improving legislative acts and norms ensuring the protection of living nature, ecologizing agriculture and industry, ensuring effective management and administration, and developing public movements to protect living nature.

All groups of countries have similar trends in the development of the organizational and legal aspects of the protection of living nature. These include: (1) the transition from government management of nature-conservation territories alone to centralized management of all spheres of nature protection, including the protection of the vegetable and animal kingdoms; (2) the growing development of a network of protected territories (in percent of the state's territory); (3) the transition from general environment protection legislation to detailed legislation dealing with individual living natural resources. This trend is borne out, for instance, by the fact that in the period between 1965 and 1975 the OECD members (the 24 largest capitalist nations) adopted 3.5 times more legislative acts for the protection of the environment than in the preceding decade (Martynov and Novikov 1981).

The spread of public movements to protect the environment, living nature in particular, warrants special attention. In socialist countries, national nature protection societies are among their largest social organizations. In the RSFSR (one of the Soviet Union's constituent republics), for instance, the All-Russia

Nature Protection Society had some 36 million members in 1984. Similar processes are also under way in developed capitalist countries. In 1980, the Greenpeace movement, which has branches in many countries, numbered over 3.5 million members in the United States alone. In the 1970's, over 3000 public ecological organizations functioned in that country. In the late 1970's and early 1980's, environmentalists, such as the Greens, achieved a marked and stable political success in a number of West European countries (the FRG, France and elsewhere). Following this, all leading political parties in developed capitalist countries began to pay much more attention to ecological problems. But the socialist and capitalist countries have common problems as regards the protection of living nature. These are: a sharp decrease in the number of some species of animals and plants, an increased number of species which have disappeared from the fauna and flora of a particular region, the intrusion of a great number of alien species (as a result of intentional or unintentional acclimatization or introduction), and the insularization of territories where wildlife is still preserved. All this is caused both by the development of industry, agriculture, energetics and transport, and by the increasing global pollution of the environment.

Both capitalist and socialist countries must develop their networks of protected territories further in order to preserve the rich variety of animals and plants, and also to maintain the ecological balance in various regions. In a longer perspective, specially protected natural territories, together with territories with a special regime of nature utilization, should account for at least 12–14% of the territory of each large region in a country. The centralized management of environmental protection measures on a nation-wide scale is to be strengthened everywhere, and the role of the long-term planning of nation-wide measures to protect plants, animals, landscapes and large ecosystems must be increased. This has been called by some the "socialization of nature" (Saint Mark 1971). Alongside stronger centralized management, it is also feasible to strengthen both government and public on-site inspection bodies.

The solution of these problems in socialist countries differs drastically from that in capitalist states because they emerge and are dealt with on the basis of state or social ownership of the objects of living nature and the principle of planned development of the national economy, including the protection of the animal and vegetable kingdoms.

Of course, in socialist countries, too, not all problems of the protection of living nature can be resolved easily and efficiently. Departmentalism in the utilization of natural resources, insufficient ecological orientation and lack of information leading to planning mistakes in the use of resources and the development of regions, and major difficulties in ecological prognostication exist objectively and hinder the manifestation of society's efforts as regards the protection of living nature.

The realization that the destruction of nature caused by the uncontrolled exploitation of natural resources may undermine the foundation of their economy compels developed capitalist countries, also, to search actively for ways of improving the protection of nature.

One such way is state-monopoly regulation. National and regional plans or strategies for environmental protection for 5–10 years have been developed in many countries, including the USA, Japan, Sweden, Britain, and the FRG.

Another way is to augment the influence of the public, to ensure the rational utilization of nature, by emphasizing not only economic, but also social benefits.

The situation with regard to the protection of living nature in the developing countries differs markedly from that in the socialist and developed capitalist states. The difference stems from (1) the developing countries' desire to ensure rapid independent economic development (which inevitably involves intensive utilization of their natural resources, and (2) the increasing export of "dirty" enterprises from developed capitalist countries to regions with cheap labour and low expenses for preserving and restoring the environment disrupted by industry. Both problems have not only purely ecological, but also social importance, and their full analysis goes beyond the framework of this book. Mankind, however, is not indifferent to the practical organization of the utilization of, say, the Amazon forests which serve as an absorber of CO_2, a source of oxygen and a climate regulation factor and ensure the preservation of many hundred thousand species of animals and plants which may disappear in the case of uncontrolled felling. This is taken into account by some leading capitalist countries. Specifically, the US environment protection programme Global 2000 says that it might be feasible to compensate countries possessing tropical forests for their cessation of felling these unique natural biomes.

For economically developed countries and regions, the most urgent environmental problem is to halt pollution and restore the quality of the natural environment, whereas for developing countries the main task is to preserve the existing ecological parameters. Many protected territories in the developing world are now in jeopardy. In 1984, out of the 43 most endangered protected territories in the world, 34 are in developing countries (Thacher 1984).

Lavish living nature is an ecological and geographical characteristic feature of many developing countries. The development of international tourism may be a form of economically profitable exploitation of this lavish nature.

Because people in the developing world live mostly in rural areas, the correct orientation of economic activities in the countryside takes on special importance. Particularly, in certain localities it would be feasible to organize ecologically safe pasturable cattle-breeding using the resources of wild flora and fauna. But a long-term solution of these problems calls for considerable investment which, as a rule, cannot be made by these countries without outside assistance.

8 Problems of Prognostication

According to the analysis presented in the previous chapters, the majority of conservation problems may be divided into several groups associated with the main levels of living matter organization. It appears convenient to discuss problems of prediction and forecast in the field of conservation issues on the basis of the same approach using this concept of the levels.

8.1 Molecular Level

Many works are devoted to the study of structure-activity relationships (SAR) of pollutants as a potential basis for the prediction of their ecotoxicological and toxocological behaviour (e.g. see Lipnick 1985).

The log of the 1-octanol/water partition coefficient (K_{ow}) is often considered as a predictor of aquatic toxicity for organic pollutants. However, at least four serious restrictions do not permit using this predictor in practice effectively (Laughlin and Linden 1985; Schultz, Applehaus 1985; Neumann et al. 1987):

1. This predictor works only within a certain chemical group of previously selected structurally related compounds. Criteria which could be used a priori for lumping and splitting pollutants into toxicologically meaningful groups are far from being clear and universal;

2. The correct K_{ow} value in real ecosystems is a problem. Taking tributyltin as an example, estimates vary between 200 and 7000 in different waters, depending on salinity;

3. The ecotoxicological danger of a pollutant depends on its bioaccumulation factor (BAF). Methods of theoretical prediction of BAF are inadequate: thus, for tributyltin in seawater, some calculations of BAF gave the value 473, but no empirical data for tributyltin BAF are as low as 473;

4. The ecotoxicological properties and toxicity of any pollutant are at least in part determined by its biotransformation, which we are often not able to predict quantitatively.

Attempts to predict the ecotoxicological properties of xenobiotics on the basis of K_{ow} are relevant to aquatic ecosystems and water pollutants. For pollutants reaching terrestrial organisms through the air or trophic chains, the methods of prediction of their properties are even less elaborated. We may "predict" only approximately the presence of toxicity for compounds of types which are well known to be toxic: halogenated hydrocarbons, pesticides, organometallic and heavy metal compounds. But to give a quantitative prediction of toxicity of really new compounds, especially of new types, is practically impossible.

The difficulties in the prediction of toxicity of xenobiotics are partly connected with the fact that diet, age, hormonal changes, environmental factors, drug-drug or xenobiotic-xenobiotic interactions may alter rates and routes of metabolites of even previously well-studied xenobiotics.

Predictions on biotransformation and biodegradation of pollutants are extremely difficult owing to the limited value of even existing data on the biodegradation of particular xenobiotics, since numerous discrepancies exist between the result found in several screening biodegradation methods, as has been shown, for example, for aromatic compounds and detergents. Thus, both types of test that were used to assess biodegradation potential (Howard and Banerjee 1984) — screening methods and grab sample tests — often do not give information adequate for predictions.

Some other data important for the problem of prediction of ecotoxicological behaviour of pollutants are considered in Chapter 2.2.

8.2 Onthogenetic Level

It has been shown by many authors that it is often impossible to extrapolate data about the toxicity of a certain compound or of a class of compounds from one species to another; moreover, in the majority of cases it is also impossible to predict quantitatively the effects of a certain pollutant on a given species by using information about the effects of other chemically related compounds on the same organism. This has been demonstrated by numerous investigators in studies of all the main classes of pollutants: pesticides, organometallic compounds, heavy metals, detergents.

This conclusion has been confirmed in many recent works and in our studies also: e.g. in the study of anionic detergents, SDS and Sulfonol, on the growth of seedlings of *Sinapis alba* and algae *Scenedesmus quadricauda* (Gorjunova and Ostroumov 1986). Sublethal concentrations of SDS inhibited the seedlings but stimulated the growth of algae. However, an attempt to use this data for the prediction of effects of another anionic detergent, Sulfonol, were unsuccessful. Our experiments have shown that sublethal concentrations of Sulfonol were more toxic for growth of algae than for the seedlings.

There are practically no effective and universal criteria suitable for prediction a priori of whether a certain compound will exhibit teratogenic properties and if so, in what concentrations.

In order to try to predict effects of pollutants on onthogenesis and survival under natural conditions, it is often essential to be able to predict the value of BAF; this is in most cases impossible, as was stressed in the previous section of this chapter.

The inability to predict the effects of a given substance on onthogenesis and the survival of an organism of a certain species on the basis of the previously known level of its toxicity for another species is at least in part connected with interspecific differences in activities of key enzymes that are involved in the biotransformation of xenobiotics; their differences even in the case of different species of livestock may be several hundred- to thousandfold, as has been demonstrated for cytochrome P-450-dependent monooxygenases and for several transferases (Watkins and Klassen 1986).

8.3 Populations and Species Level

Population genetics permits the prediction of a rapid decrease in genetic variability (and hence ecological stability) of the isolated population with low effective population size (N_e) which can be as low as a tenth of the actual number of animals in the population. Thus, the genetic variability of a population with $N_e = 25$ will decrease more than 50% within 40 generations. This predicts an extraordinarily low stability of populations of several tens or even hundreds of animals; unfortunately this situation is not uncommon — examples of such populations are given in Chapter 4 in the sections on extinction of birds and animals.

Among the problems of ecological forecasts are those associated with the problem of estimating the speed of the evolutionary process in populations. For instance, this problem is important in the case of anthropogenic evolution of resistance to pesticides in pest population. The appraisal of the speed of evolution of resistance may help in the attempt to make an ecologically and economically important prognosis.

Recently the process has been very rapid: within the 10 years between 1970 and 1980, the number of harmful resistant insects and mites nearly doubled (from 224 to 428), within the 20–23 years between 1960 and 1980–1983, the number of taxons of fungi and bacteria resistant to pesticides increased 3.5-fold (from 20 to 150).

Resistance to particular pesticides may develop within a few seasons: thus, within 10 years (1970–1980) the number of insect species resistant to carbamates increased 17-fold, within 18 years (1956–1974) the number of disease vector-insect species resistant to organophosphate compounds increased 27-fold (Dover and Croft 1985); within 3–4 years (1976–1980) the number of pests resistant to pyrethroids advanced ca. threefold (from 6 species to 17).

The rate of evolution of resistance seems to outrun the rate of developing a new chemical, which takes from 8–10 years from discovery to marketing and costs US $20–45 million (Dover and Croft 1985).

The speed of resistance development was quantitatively measured in house mouse population which increased its tolerance to DDT twofold over ten generations of exposure. However, in insects the process may be even faster, and increases in level of resistance as high as 25,000-fold have been recorded (Pimentel and Edwards 1982).

We could assume that the constant rate of appearance of resistance in insects and mites is about 200 species per 10 years (as it was between 1970 and 1980). Assuming additionally that (1) the total number of main pest species is about 2000 (Pimentel and Edwards 1982) and (2) at present ca. 500 species are already resistant, we may project that within 75 years almost all the main pest species may become resistant, if the present level of pesticide application continues. However, in the case of a twofold increase — which is expected before the year 2000 — the triumph of the pests will overrun us twice as rapidly.

At the species level, another important projection deals with the total process of extinction of species. We may firmly predict that the extinction of species will continue and "hundreds of thousands of species can be expected to be lost by the year 2000". However, this prediction is based upon the suggestion that a certain percentage of total biospheric amount of species will be lost by the year 2000, and that there are a few million species on the Earth in toto. But recent estimates point to a larger world biota — 10 million species or even more (Wilson 1986). Assuming that an important part of them is harboured in tropical forests, we may conclude that the extinction association with ecosystem destruction will be of even greater magnitude, i.e. *more* than hundreds of thousands of species by the year 2000.

Different scenarios of probable extinction of species are given: (1) assuming that tropical forests harbour of 750,000–2,500,000 species (as has been done in Global 2000), (2) assuming that tropical forests contain 1,500,000–5,000,000 species.

The examples of the forecasts discussed above in this section are all connected with different negative trends. It is much more difficult to make forecasts of a positive sort that predict the future for normally existing or flourishing populations; it may often be even impossible for real field conditions.

One of the few types of realistic forecasts of a positive sort is the demographic forecasting of the behaviour of closely managed populations (Soule and Wilcox 1980). It should be stressed that in nature this is a rare situation of isolated and well-studied population where many parameters are known: age and sex structure, fecundity and mortality for every age class. This type of prognosis deals with an evolutionarily very short period of time, and does not take into consideration the processes of microevolution.

8.4 Ecosystem and Biosphere Levels

Many attempts to predict the behaviour of ecosystems and of the biosphere on the basis of mathematical models exist, many of them of considerable value and importance. However, it may be noted that a significant portion of mathematical models of ecosystems has no real predictive value for field conditions.

It is practically unrealistic to construct a model that predicts behaviour of the major part of hundreds of individual species of a certain natural ecosystem. One of the obstacles that seems to be very serious is the difficulty of taking into account all or at least the majority of potential links between species. If a given ecosystem has N species, then the amount of possible links between species is $N(N-1)/2$. This means that in the case of an ecosystem composed of 2000 species (such figures have been established, e.g. for some lakes) we must consider $(2000 \times 1999)/2 = 1,999,000$ links between species. The complete analysis of this amount of links is hardly realistic.

Another difficulty in the prediction of behaviour of ecosystems is connected with the lability of species composition in ecosystems. Even the set of species (in addition to the relative abundancies of species) is an inconstant parameter, because the possibility of introduction of a new species is never excluded, and any introduced species may carry with it a whole set of associated species of parasites and pathogens (a sort of ecological cluster of species). The appearance of a new species in the ecosystem under consideration in terms of prognostication may make the predictive value of the old prognosis equal to zero. It is known from the analysis of consequences of the introduction of mammals, hydrobionts and plants that the results of such introductions are usually unpredictable.

It is important to stress that the dynamics of an ecosystem with its productivity and flows of energy and matter between its compartments, as well as the real dynamics of populations are often determined by a group of practically unpredictable factors, i.e. fluctuations of weather and climatic conditions. Many examples of the study of flows of matter from one compartment to others during several vegetation seasons demonstrated the existence of 100–200% changes annually in these processes. It should be said that often these changes are not caused by succession or by other predictable factors.

It has been firmly established that in many ecosystems, fluctuating changes of several dominating species occur from one year to another. The unpredictable sort of annual changes is often demonstrated by plant communities where at least four types of fluctuations were described: (1) fluctuations of numbers of individuals in populations connected with changes of ecotopes (ecotopic fluctuations), (2) those connected with the action of herbivores and digging animals (zoogenic fluctuations), (3) those connected with life cycles of plants and/or with unequal seed production or vegetative reproduction from year to year (phytocyclic fluctuations), (4) phytoparasitic fluctuations (Rabotnov 1983). Analogous unpredictable fluctuations of populations were also found in animals.

Important experience was obtained when in the 1930's–1950's in the USSR a number of giant man-made reservoires were constructed. In those days the most prominent hydrobiologists and ichtyologists had prepared ecological forecasts. In retrospect, it is clear that some important points of the forecasts were wrong. Thus, nobody predicted a large-scale dissemination and translocation of species of hydrobionts from North to South and from South to North. Unsuccessful were also some predictions concerning the formation of benthic communities and of ichtyofauna. Thus, the real biomass of zoobenthos in several man-made reservoirs (e.g. Tsimljanskoe, Volzhskoe, Rybinskoe) was 5–15-fold smaller than the figures predicted. The fish productivity was 2–8-fold lower than predicted (Nikolaev 1980).

It seems probable that the disappearance of some species in the ecosystem may trigger a whole chain of events of disappearances of other species which depend on the first species. This may take place in the case of the obligate and specific dependency of one species on the other, e.g. in the case of the existence of a sort of association of species where at least one of them is inevitable for the survival of the other species. Such ecological clusters of species have been insufficiently studied.

Examples of reliable predictions concerning ecosystems are the forecasts based on the multiannual trend of destructive processes in several types of ecosystems. This is true, for example, for tropical forest ecosystems (Chap. 5) and also for agroecosystems.

In the USA, agricultural ecosystems with corn (maize), millet and cottons are predicted to decrease their production potential, especially when located on sloping lands, owing to erosion which causes a loss of as much as 20 tons of soil per acre per year on gentle slopes. Of 283 US farms surveyed, 84% had annual soil losses in excess of 5 t per acre, which is the maximum that can be sustained annually without harming productivity (The Global 2000, 1980).

Urban development between 1978 and 2000 is expected to occupy more than 25 million ha of cropland (2–2.5% of the world crop base) which even under average productivity may feed some 84 million people. The real loss will probably be even higher, since India alone projects non-agricultural land use to expand by 9.8 million ha between 1970 and 2000 (G 2000).

The inclusion of virgin lands into cultivation will induce the deterioration of their ecosystems: e.g. the soil organic matter declines under cultivation to an equilibrium value of 40–60% of the original content.

In many agricultural ecosystems we may predict soil deterioration with a concomitant decrease of potential for primary productivity (PP) and damage to soil biota as well as other negative trends including: (1) erosion, (2) deprivation of organic matter, (3) build-up of toxic salts and chemicals, (4) decay of the porous soil structure. However, it is difficult to predict quantitatively the course of these processes in particular ecosystems.

In the case of some agricultural ecosystems, forecasts of decline in the PP between 1985 and 2000 are based on depletion of the water supply: e.g. on the

High Plains of Texas due to depletion of the Ogallala Aquifer; in western Kansas and Nebraska owing to their water table drop (Moore 1985).

In conclusion, some estimates, considering all factors including desertification, waterlogging, etc, predict up to 30% loss of arable lands within the next decades.

Difficulties of quantitative forecast of processes in ecosystems are often connected with: (1) the delayed and buffered character of response of ecosystems to external influences and (2) the contra-intuitive behaviour of ecosystems.

Regarding successful forecasts at the ecosystem level, the following three points should be stressed.

First, many of them consider relatively simple and permanently stable ecosystems that are rare in nature.

Second, many successful forecasts relate to extreme or pathological situations: e.g. predictions of species extinction ("faunal collapse") as a result of insularization, of extinction due to inbreeding depression in small captive populations in zoos or in diminished natural populations (Soule and Wilcox 1980).

Third, the creation of a detailed and adequate model of a real ecosystem is extremely expensive. One models of a prairie ecosystem in the USA cost more than US $10 million. The final price of collecting information and modelling may exceed the benefit of forecasting.

On attempting to make a prognosis of the behaviour of systems at the biosphere level, all these difficulties are additive; moreover, additional obstacles are: the large-scale character of processes, high level of inertia, the difficulty of measuring the changes in biospheric cycles of elements, the low precision in recording the state of biospheric resources.

Some existing biospheric forecasts predict the following.

1. Under continuation of the present trends of population growth, increase in agricultural production associated with intensification of agriculture, industrial production growth, a further increment of pollution of biosphere will occur.

2. Inevitably predicted are further growth of consumption of resources of the biosphere, decrease of some natural resources especially in per capita calculation, as well as the destruction of the most exploited ecosystems and habitats.

3. Essential consequence of the pollution of the atmosphere will be a warming of the climate, especially in boreal and subboreal regions, changes of water transpiration by plants, shifts of borders of climatic and biogeographic zones, a new pattern of distribution of ecosystems and species over the land surface, extinction of a number of protected species and disappearance of some ecosystems in reserves and national parks, translocation of zones favourable for agriculture over the Earth's surface.

Impressively concrete and realistic predictions have been obtained due to the modelling of the biospheric consequences of a nuclear war.

8.5 Forecasts of the Ecological Consequences of a Nuclear War

Modelling of the ecological effects produced by a nuclear war has shown that even a "small" nuclear war associated with the use of 100 Megaton (Mt), i.e. less than 0.8% of the world nuclear arsenal in 1983, will create substantial amounts of smoke and air-borne particles; these amounts will lead to global climatic changes, called "nuclear winter" (or in another independent work "nuclear fall"). On the use 5000 Mt — less than 50% of the 1983 world nuclear arsenal — fires will produce about 225 million t of smoke and other particles. The intensity of solar illumination will decrease and attain only a few percent of the normal level. Moreover, in the case of a 10,000 Mt war the level of illumination will drop to 0.1% of the normal level (Turco et al. 1983; Svirezhev et al. 1985; Moiseev et al. 1985 and others).

The air temperature over the Earth's surface will decrease; the amplitude of this decrease may reach 23°C after a 5000 Mt war and 50°C after a 10,000 Mt war. Other models predict higher or lower amplitudes of temperature decrease. However, the difference between nuclear winter and nuclear fall is irrelevant: even a 4°C drop in temperature during the vegetation season may be enough to destroy the whole regional yield of wheat and barley. Moreover, the empirical basis of the model predicting a nuclear winter is well documented and leaves practically no room for substantial mistakes in forecasts (e.g. Moiseev et al. 1985). Hence predictions of a nuclear winter seem to be reasonable.

The radiation dose to which inhabitants of 30% of the temperate area (between latitude 30°N and 60°N) will be submitted in the case of a 5000 Mt war is more than 250 rad over several months; inhabitants of 50% of this area will receive more than 100 rad. These doses are very dangerous.

War-induced decrease of ozone in the atmosphere will increase UV-radiation, which in turn will cause detrimental effects for terrestrial ecosystems.

Ecological modelling even in the case of a "small" nuclear conflict predicts the disappearing of forests, increase of desertification, floods, disorganization of agriculture, expansion of ruderal vegetation and large-scale destruction of numerous ecosystems (Ehrlich 1983; Svirezhev et al. 1985; Grime 1986). In the case of a large-scale war, it seems inevitable to predict not only a nuclear winter but, taking into account numerous ramifications, nuclear annihilation of the biosphere in its contemporary form.

By which ways (if any) can we evade the present situation? The following part of the book is devoted to this theme.

Part II

Prospects for the Protection of Living Nature

Part I of this book discusses some major problems of the protection of living nature. Part II is an attempt to offer solutions to these problems.

Our basic premise is that the conservation of rich animate nature is a sine qua non for the normal functioning of the biosphere at large and its individual regions. This makes it the most important prerequisite for the development of mankind, rather than an impediment. There is no single sphere of society's life not affected by these problems, be it agriculture or energetics, medicine or politics and law.

Chapter 10 deals with need for "ecologization" trends in society's activities. It is the "ecologization" of the utilization of natural resources and virtually of all the aspects of society's life (this process gaining momentum over the past 15–20 years) which may give ground for some hope as regards finding solutions to the major problems of the protection of living nature as a whole. Chapter 11 considers special ways of resolving the chief problem, viz., the preservation of life in all its variety on our planet. But before that we will discuss three other global problems closely connected with nature protection (Chap. 9).

9 Protection of Living Nature and its Connection with Other Global Problems

All today's global problems closely intertwine and exacerbate each other. We will discuss here only three such problems, whose solution is a must, because otherwise progress in animate nature protection is unthinkable. First comes the problem of averting nuclear war.

9.1 "Nuclear Winter" and Disarmament

If a nuclear conflict flares up, any activity to protect life automatically becomes senseless.

Clouds of dust, smoke and soot from fires caused by nuclear explosions will be blown out into the stratosphere well before the radioactive fallout destroys the genetic pool of higher organisms (which may happen months or even years later). This will immediately deprive the planet surface of the usual quantity of solar energy; the air temperature will drop (e.g. Chown 1986), a "nuclear winter" will set in (e.g. Greene et al. 1985). Nuclear explosions will inevitably lead to other harmful consequences, such as changes in the ozone content in the atmosphere and, as a result, changed levels of UV radiation. If the ozone content drops 50% (which is quite possible according to some estimates), UV radiation intensity will increase by scores of times. Several months later, therefore, when the explosion-caused dust and soot have settled, intensive UV radiation, formerly restrained by the ozone screen in the atmosphere, will prove ruinous for every living creature (e.g. Ehrlich 1983).

But even without nuclear war, mankind cannot, given the present rate of the arms race, provide sufficient funds even to maintain the present level of environmental protection, low as it is. In many countries, military expenditure is already tens and hundreds of times higher than expenses for environmental protection. Only putting a halt to the reckless arms race can provide the financial and material resources so badly needed to improve the environment (and save many species from extinction).

The arms race leads to the destruction and exhaustion of mankind's natural resource base, environmental pollution, destruction of biogeocenoses over vast areas, hypertrophied development of industries not vital for man, and excessive urbanization. The very need to maintain an acceptable quality level of the environment and, correspondingly, a certain sufficiently high level of living nature protection is certainly an essential factor today for maintaining peace on the Earth and developing worldwide cooperation.

9.2 Demographic Problem and Need for Stabilization

The progress of mankind and the biosphere as a whole are confronted with another global problem, viz., the population explosion (e.g Ehrlich et al. 1977; Peterson 1984). The share of mankind and the biomass of man's domestic animals accounted for 15–20% of the total biomass of land animals in 1980, and it continues to grow.

In 1985, the number of people living on the planet totalled some 4.8 billion, of whom one billion were undernourished or starving.

In the period from 1970 to 1986, the world's relative population growth rate has markedly decreased from 1.9 to 1.62%. In Great Britain, West Germany and some other European countries the population has begun to decrease, although the Earth's population in general may stabilize as late as the early 21st century. Even if the Earth proves capable of feeding that number of people, the biosphere will hardly be able to withstand the ever-increasing torrent of various wastes and pollutants (chemical, heat, optical, etc.). The UN Conferences on Population held in Colombo (1979) and Mexico (1984) persistently urged the world community to drastically cut the population growth rate, especially in the developing countries, which moving to effect economic and social reforms. Population bomb (Ehrlich 1980) is a danger for life-support systems of the biosphere.

9.3 Reduction of the Biosphere Pollution Rate

In some industrial regions of the world, the growing rate of biosphere pollution with certain types of pollutants seems to have passed its peak and now shows a stable downward trend. This has not happened all by itself, but rather as a result of persistent, sometimes unprecedented efforts by individual states and groups of countries. Sewage water in many big cities in Europe, North America and Asia is now treated more thoroughly and discharged into reservoirs in smaller quantities. Biota has been restored in Lake Erie, one of the Great Lakes, and salmon has reappeared in the Thames after the many years of absence. The number of fish species in the Seine has grown from 8 to 32 over the past 15 years (Fig. 9.1).

The gas and dust pollution of the air has considerably decreased in many regions. Nearly one-third of all metal in the world is now extracted from used industrial hardware. Despite the marked over-all increase in oil transportation by sea and in shelf oil production, the total volume of oil spills in the ocean from anthropogenic sources seems to have decreased.

Water treatment facilities built in the USSR in 1983–1984 with an aggregate capacity exceeding 17 million m³ of sewage water a day have made it possible to reduce the volume of polluted drains into reservoirs by 2.9 km³ annually. In the period between 1981 and 1985, industrial output in the USSR increased 24%, yet the industrial use of water remained practically the same. This was due to the application of closed water cycles and consistent water reuse. By 1984, the volume of reused and recycled water increased by 36 km³ as compared with 1980. The systems of reused and recycled water accounted for 69% of the industrial water requirements in the USSR in 1984. Between 1975 and 1983, the total amount of harmful substances blown out into the air by stationary pollution sources in the USSR decreased by almost 9 million t despite the markedly increased volume of industrial output.

Philosophers say that freedom is a conceived necessity. The conceived necessity to save our forests from death caused by acid rains has resulted in an internationally agreed plan of action to reduce SO_x and NO_x exhausts by all European states irrespective of their political system — from monarchies to socialist community countries. One need not be a prophet to say that many similar plans and agreements will be reached and implemented in the very near future. The Earth has proved to be a small home, whose inhabitants, whether they like it or not, must comply with the rules of social intercourse to maintain a decent life. These rules will grow ever more rigid in proportion to the growth of man's

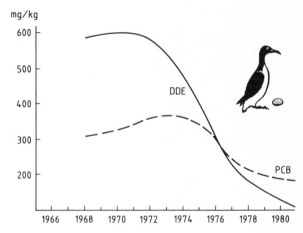

Fig. 9.1. Dynamics of DDT/DDE and PCB contents in the eggs of longirostal plover (*Uria aalge*) in the Central Baltic. (After Olsson 1977)

technological potential. This begins to bear out the proposition of V.I. Vernadsky (1977), one of the most prominent contemporary natural scientists, who said, "The life of mankind, however varied, has become indivisible, integral. . . . This is the beginning of a spontaneous movement, a natural phenomenon which cannot be stopped by the fortuities of human history. For the first time ever, man has really understood that he is an inhabitant of the planet, that he can, and must, think and act in a new aspect — not only in the aspect of individual personality, family or tribe, state or state alliances, but also in the planetary aspect. . ."

10 Need for Ecologization of Society's Activities

In the 20th century, industrial production, commodity consumption and other development indices are going up the exponential curve. Neither the Earth's biosphere, nor economy, nor man as a biological species will be able to sustain this growth for long. As man's economic activities are growing in intensity, they are bound to blend ever more organically with the biosphere. There is no alternative to this process of ecologization, because otherwise mankind will not be able to survive. Let us consider some of the needs and ways of ecologization of society's activities that are already in evidence.

10.1 Need for Ecologization of Agriculture

The ecologization of agriculture involves a number of issues, of which we will discuss the following: (1) transition from chemical to biological methods of crop protection; (2) transition from monocultures to polycultures; (3) ecological organization of agricultural landscapes; (4) transition from extensive to intensive methods of agriculture.

10.1.1 Transition from the Chemical to Biological Methods of Crop Protection

The drawbacks of the chemical methods are evident: lower efficiency caused by resistance (e.g. Agarwal 1979) and high costs, the extinction of many other living organisms and harm to Man's health (e.g. Balk, Kieman 1984). But there is a way out: chemicals should be replaced by various biological methods of protection (Pimentel and Pimentel 1980). This involves measures to purposefully maintain certain species (or their inclusion in agrocenoses), which would keep down the numbers of other species undesirable in an agricultural landscape (Fig. 10.1).

The control of prickly-pear cactus thickets and rabbits in Australia was a graphic illustration of the effectiveness of biological methods. In 1860–1870, two

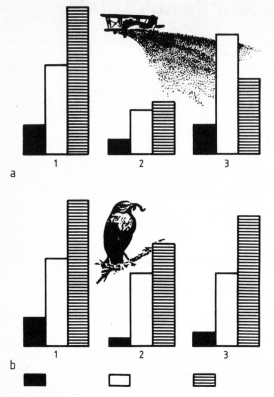

Fig. 10.1a,b. Compared efficiency of chemical (**a**) and biological (**b**) methods of plant protection as applied to garden species in Latvia (USSR). The use of chemical pest-killers does not result in a stable decrease of entomophagous species and does not suppress pests adequately. *1* prior to the use of pesticides; *2* immediately following their use; *3* one year after their use. (Yablokov and Ostroumov 1983)

species of such cactuses were brought to Australia from the United States to form live fences around pastures. By 1887, the cactuses spread so much that they became a dangerous weed which had to be fought vigorously. In 1920, prickly-pear cactus thickets occupied 30 million ha and were spreading at a speed of 0.5 million ha a year. The problem was solved as late as 1939, when a small moth (*Castoblastis cactorum*) was brought in and multipled. Its caterpillars feed on the pulp of cactus stems. By now, only scattered plots with prickly-pear cactuses are preserved in Australia, together with a small population of the moth living on these plants. A new ecological balance has thus set in. A similar process is characteristic of another anthropogenous scourge of Australia's nature and agriculture – the European rabbit, also brought there in the 19th century and then controlled by mixomatosis.

The poisoning of coyotes in North America, or wolves in Canada or Kazakhstan, is fraught with dangerous shifts in the ecological balance over vast territories, to say nothing of the possible death of many useful animals. The all-out extermination of wolves is also inefficient. It has been proved in many studies that the wolves' place is taken either by dog-wolf hybrids or wild dogs which are much more difficult to control. There is a biological method of protection, which is simple and has been known to sheep-breeders for hundreds of years: the use of sheep-dogs specially bred by our ancestors to guard sheep.

It should be emphasized that biological methods of protection differ in principle from chemical ones in that the ultimate goal in this case is not the full extermination of undesirable species, but only reduction of their numbers to a safe level (e.g. Browning 1974).

In future, the wide use of chemical plant protection methods (Heinisch et al. 1976) should, of course, be given up, for it is an ecologically "dirty" way of boosting the efficiency of agriculture. Instead, use is to be made of the ecologically "clean" biological methods, as well as natural regulators as pheromones, repellents, antifeedants, anti-ovipositants etc. (Ostroumov 1986; Grainge, Ahmed 1988).

10.1.2 From Monocultures to Polycultures

The purposeful complication of agrocenoses is one of the most interesting and ecologically promising trends in transforming agriculture. Some complex biogeocenoses, such as forests, feather-grass steppes and shallow-water ecosystems, are most productive as regards the level of solar utilization on Earth. A switch-over to polycultures promises less vulnerability to insects and also to fungus and virus diseases: in complex ecosystems the interconnection of species is such that sharp increases in the numbers of one species are quickly neutralized by feed-back mechanisms.

Experiments have shown that three-species agrophytocenoses, such as pea-mustard-sunflower and vetch-mustard-sunflower, give higher and more stable yields of green fodder. The weed-growth level is reduced 200–300%, which makes the use of any chemical weed killers redundant.

The fields of the future may well be multi-layered and produce yields for most of the year: one crop will ripen very early, another somewhat later, and so on. A complex combination of animals and plants is quite possible in such future agricultural complexes. Rice fields, where carps are bred simultaneously, are forerunners of such complex.

The organization of a polyculture may produce another effect, important from the viewpoint of the protection of living nature: a great number of species and forms, once living in "wild" nature alone, will prove to be the object of man's economic activities and will thus be saved from extinction. This holds good for plants and animals equally.

10.1.3 Ecologization of the Landscape

A vast field stretching to the horizon with no shrubs or trees in the vicinity was once considered an ideal agricultural landscape. But that time is now past. A monotonous agricultural landscape was but one stage in land development. This kind of landscape is good for mechanized cultivation, but it runs counter to the ecological need to maintain variety as a means for stabilizing ecosystems. A homogenous landscape sharply reduces the biotic variety. In Middlesex (Britain) alone, 78 plant species disappeared as a result of this process in the period between 1869 and 1975.

A certain diversity of the environment is required for the existence of most types of wild animals and plants that stabilize agrosystems. Islands of virgin soil, such as untilled spots of land with trees in the centre, narrow gullies filled with water, heaps of boulders, thickset hedges, strips of fallow land and roadsides covered with grass and brushwood are all not only life-saving spots for many species of animals, but also a bridge-head for an offensive to control species damaging the harvest. These are nesting places for many birds, including partridges and quails, and also for the Hymenoptera — the chief pollinators of flowering plants on our fields and pastures (80% of the Earth's plants need to be insect-pollinated).

It is common knowledge that during the reproduction season insectivorous birds consume immense quantities of insects, thus being a powerful preventive to their mass propagation. Small passerines make hundreds of sorties a day to catch their prey, their flight radius not exceeding 300–400 m from the nest. This is why birds usually control only a small strip of the field 100–200 m off the forest edge.

The attraction of birds to fields, the purposeful formation of aviafauna, teriofauna (the use of foxes, weasels and polecats to control mice rodents, rather than poisons), herpetofauna (certain reptiles are even more efficient insect "suppressors" than insectivorous birds), and the introduction of a set of insects useful for agriculture make it possible fully to "organize ecologically" the agricultural landscape and reduce or even exclude the need for chemical herbicides.

10.1.4 From Extensive to Ecologically Safe Intensive Methods of Agriculture

Agriculture has been developing extensively since its emergence 15,000–10,000 years ago. Its intensive development began comparatively recently. This has resulted in the shrinkage of arable land in some developed countries, which began in the 1960's. Simultaneously, their farm produce output increased. The scale of such intensification is impressive. In Britain, for example, crops increased 100% in the period between 1946 and 1966 (23), whereas the average yield of cereals in Holland grew as much as 900% between the 17th and mid-20th centuries. Some plots produce from 9 to 10 mt of cereals per ha, which is considered close to the theoretical limit. Nearly 300 kg of nitrogenous, phosphorus and potassic fertilizers are applied to each hectare, which gives rise to another set of specific environmental problems.

The extension of hothouse farming is the next step towards intensifying agriculture and making low-yielding land areas available for restoring the disturbed ecological balance. Estimates and diagrams have already been made for a more rational use of solar energy in huge hothouses stretching over tens of hectares. In them, vegetation continues throughout the year and the yield may exceed that of an open field several times over.

The next step in that direction is the establishment of phytodromes. Their designs, based on real production experiments, have already been worked out. Plants are grown in movable cassettes which are moved forward depending on the duration of the vegetation period of each plant species and the lighting regime. Sowing and harvesting are performed every day on each automatic line. This multi-tier conveyor production system makes it possible to assimilate solar energy with greater efficiency and increase the yield scores of times. Some fantastic (yet based on real figures) calculations have been made, showing that a 300×300 km phytodrome in the Kara-Kum desert in Kazakhstan would be able to supply food for 300 million people.

The ecologically safe intensification of agriculture will possibly help to free vast areas needed for the existence and development of the animal and vegetable kingdoms on our planet. In the future society, the now routine forms of agriculture will, possibly, yield place to polycultures combined with phytodromes. In case of non-ecologization the agriculture may "swallow up" the last remnants of virgin soil preserving the wild genetic pool.

10.2 Need for Ecologization of the Economic Forms of Biosphere Utilization

From time immemorial, Man has been taking the animals and plants he needed away from the biosphere, giving nothing in exchange. But that time has now gone forever: it is foolhardy to take something away from nature with no provision for restoring and maintaining the numbers of any object of Man's economic activities. Moreover, once Man's responsibility for restoring a resource could be confined to a regulation system (like a total ban on hunting or fishing), whereas now this rule proves insufficient in most cases.

Let us consider the main trends in the ecologization of Man's economic activities as applied to living nature.

10.2.1 Hunting

Some applied disciplines deal entirely with the rational utilization of animals and plants procured by man on a large scale, their many theoretical propositions being also helpful in resolving major environmental problems. Mention should be made here of the methods developed for controlling the population of game species, regulating their reproduction and behaviour, distributing predators optimally according to biotopes, etc. (for surveys, see Dezhkin 1983; Watt 1968; Cushing 1975, and others).

Another factor is important from the viewpoint of living nature protection and is connected with the ecologization of hunting: in all cases the most productive hunting is connected with intensive anthropogenic development of territories. The population of large hoofed animals in large virgin forest areas is much less than in several smaller areas (of the same combined size), where plots of arable land are alternated with urbanized ecosystems (Fig. 10.2). In the USSR, for example, hunting is most productive in the densely populated western regions. Even in the Moscow Region, it has become necessary in recent years to shoot some 2000 elks, over 1500 wild boars and dozens of deer annually to maintain the ecological balance. Moreover, the increased beaver population has made it possible to allow, beginning in 1965, the licensed shooting of these rodents, too, once believed to be disappearing. The game yield from 1000 hectares of hunting territory amounts to more than 300 kg in Latvia, about 140 kg in Lithuania and as little as 18 kg in Kazakhstan. This factor is extremely important, given an imminent increase in the anthropogenic pressure on the Earth's biosphere.

Let us consider several examples characterizing the present day situation as regards some major groups of game animals. The saiga antelope (*Saiga tatarica*) is a clear example of how a species was saved from extinction as a result of timely hunting regulation measures, but now that nearly 5 million saigas have been shot

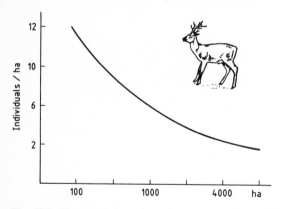

Fig. 10.2. The population of roe deer (*Capreolus capreolus*) in habitats with various levels of forestation. *Y-axis* the average number of roe deer per hectare of forest; *X-axis* the area of wooded plots. (After Bluzmo data)

in the Soviet Union over recent years, it has become clear that regulated, scientifically based hunting alone is not sufficient. What is needed is certain expenditure to keep up their habitats and migration areas in a state that ensures high numbers of species for gameshooting. The wide spread of cultivated lands in Kazakhstan requires specially organized zones for the seasonal migration of saiga herds that run into many hundred thousands of animals and makes it necessary to preserve vast semi-desert areas in intact or under-developed condition.

The elements of economic management emerge as the chief trend in the exploitation of virtually all major land game mammals in Europe, such as elk, wild boar, deer and roe. These elements include the laying-in of fodder supplies for winter, the protection of animals from predators and their large-scale transfer to localities isolated from major high-density areas as a result of industrial development.

A considerable part of game birds and some game animals in Europe are bred on special farms and then set free in hunting ranges. In France alone, 5–6 million pheasants, some 1.8 million partridges, nearly one million wild ducks and several hundred thousand wild boars were set loose in the early 1980's (Biali 1983). Indicative here was the fate of the wild turkey in the United States, which began to disappear in the early 20th century. In the 1970's, the total population of this species exceeded 1.5 million as a result of farm breeding and then setting loose or transferring to the former habitats and some other such measures. This despite the fact that intensive wild turkey hunting continued. The population of the big prairie grouse (*Tympanuchtus americanus*), which was disappearing in the United States not so long ago, came close to one million. So far there is no real danger that the wood grouse (*Tetrao urogallus*) will be exterminated as a species, although it has almost disappeared in some regions in Europe. This makes it imperative to start immediately looking for ways of farm-breeding a matrix herd. This work is being successfully carried on in the Darwinsky and Berezinsky Preserves in the Soviet Union. West Germany is already confronted with the problem of breeding the hazel-grouse (*Tetrastes bonasia*), which is still a popular game bird in the Soviet Union. All these and many other examples show that elements of economic management spread to hunting areas which have until recently been what is commonly referred to as "gifts" of nature.

The breeding of poisonous snakes, the suppliers of poison for medicine, is an example of economic management as regards reptile procurement. In 1979 alone, 5000 snakes, mostly vipers (*Vipera berus*) born in the Moscow Serpentarium, were set free. Of course, this is only the beginning of the restoration of the Soviet Union's herpetofauna: in Azerbaijan alone tens of thousands of snakes were caught annually only several years ago. Farm or semi-captive breeding will be the next step towards ecologizing the Soviet Union's "snake economy".

10.2.2 Fishing (Aquaculture)

Economic management seems to have scored the most spectacular successes in fishing. Modern Pacific salmon (*Oncorhynchus*) fishing is evidently largely based on artificially bred species. The population of sturgeons (*Acipenser güldenstädti, Huso huso, Acipenser stellatus*) in the Caspian Sea is maintained at a high production level exclusively because of artificial breeding. A white salmon subspecies (*Stenodus leucichthys*) has also been saved there thanks to artificial breeding, and its population continues to grow. Major successes have also been recorded in breeding scores of other fish species once almost completely depleted by the fishing industry. It is not fortuitous that research programmes have been developed to organize comprehensive utilization of marine sources within the framework of entire seas. Such programmes seek to keep up the population of many animal species at a high level (for example, many fish species, the crustacea and seals in the Caspian, and fish and seals in Lake Baikal).

In future, fishing must completely turn into aquaculture, whole scale and achievements are impressive even now. In Japan, for example, one hectare of plantation yields almost 500 tons of mussels. In the mid-1970's, the total annual harvest from Japan's sea farms exceeded one million tons. The scale of mariculture in other countries is also big: over 2.5 million tons of seafood is gathered annually in China, whereas worldwide aquaculture produced over 10 million tons of seafood in 1980. It is estimated that by the year 2000 the output of aquaculture, both marine and freshwater, may total 30 million tons a year, i.e. nearly 40% of today's catch of seafood in the World Ocean.

The principle of polycultures can be used on a wide scale, which is one of the advantages of sea farming. Shrimp can be bred in paddy-fields, while perch, carp, and shrimp can be grown in one and the same pond.

In fish aquaculture, feed is highly effective. From 0.9 to 1.5 kg of feed is required to produce 1 kg of trout, whereas it takes from 3 to 6 kg of fodder to produce 1 kg of pork or beef. The cost of animal protein in marine culture is 20–30 times lower.

It is believed that there will be no need to reduce agricultural territories to develop aquaculture. It is highly promising to grow marine products on rafts and in fish-wells at sea, and also to use part of the mangrove-covered territory in the tropics. Moreover, the existing 30,000–40,000 km^2 of fish-breeding ponds in the world today could be supplemented with a much larger territory if industrial waste lands are used. The utilization of the warm water of electric power stations in some cases may be promising for freshwater aquaculture. It seems better to use it for aquaculture than to discharge it to natural water bodies as thermal pollution.

More than 35 million tons of seafood out of the over 80 million produced today is used to make fertilizers and fodder for animals. This way of utilization of world fish resources is often compared with the use of high-quality food cereal to feed animals. Globally, mankind will certainly be better off if only foodstuffs are procured from the sea and the ecosystem of the World Ocean is not undermined by extracting intermediate links in food chains.

Aquaculture in a broad sense, i.e. man's active participation in the organized breeding and maintenance of water animals at a high level, reflects a trend towards the ecologized utilization of the Earth's reservoirs and allows us to be optimistic about the chances to preserve the variety of life in the World Ocean.

10.2.3 Industrial Utilization of Vegetation

As regards vegetation, economic management factors are explicitly applicable to forestry only. The share of forest restoration and forest plantation will continue to grow. In future, it will be feasible to abandon timber felling and logging in natural forests and switch over to forest plantation systems. Such systems are most effectively developed in Southern Europe. In the period between 1960 and 1980, forest restoration proceeded at a pace of 150,000 ha a year (Global 2000). In recent decades, China has been effecting a large-scale programme of forest restoration, having planted tens of millions of hectares of forest over the past 30 years. By 1990, the forest area in China, the Democratic People's Republic of Korea and South Korea is to increase by 4% as compared with 1970. Forest areas will also be expanding in Mongolia and some other countries.

Special agroforestry programmes (the integration of agriculture and forestry on forest plantations) are now being widely used in many countries. Essentially, some of these programmes boil down to developing forest and polycultures, while preserving the variety of natural vegetation in between trees. Some variants of agroforestry provide for combined growing of trees and agricultural crops and plants on the same territory, such as the "Hevea-coconut-manioc-banana-soybean-corn" system in Costa Rica. In using such systems, care should be taken of allelopathy-connected difficulties; but these are easily overcome if the problem is dealt with scientifically and the crops are selected properly. It is estimated that the total area of forest plantations in Amazonia (South America) alone will increase from 4.7 million ha in 1975 to 16.4 million in the year 2000.

The scale of forest plantation in the USSR is also large: 12 million ha in 1971–1975, 11 million in 1976–1980, and 8 million in 1981–1985. As a result, the area of forest cultures in the Soviet Union exceeded 2% of the total forest area in 1980, the percentage being much higher in the sparsely forested republics in the western part of the country (Table 10.1).

Forest restoration and forest plantation are not the only ways to introduce economic management in forestry. Forest fertilization techniques are widespread and effective forest fire fighting measures are taken, which are, of course, also elements of economic management. But such measures are not always in keeping with the ecologization of forestry. Specifically, great efforts are being made to suppress the magnitude of some particular species, which is ecologically unjustified. It has been convincingly demonstrated that phytophagous insects populate primarily weak and sick trees and are powerful natural selection agents at all stages of tree growth, thus facilitating the formation of stable plantations (Isaev et al. 1984).

Table 10.1. Area of forest cultures in the USSR and some other countries (% of total forest area)

Region	% of forest cultures
USSR	2.2
Belorussia	20.3
Estonia	23.1
Lithuania	26.4
Ukraine	35.1
Moldavia	46.0
Hungary	36.3
World	0.8 (of land surface)

Wild medicinal plants are a good example showing the feasibility of introducing economics into the process. In the USSR, for instance, the requirements of the medicinal industry in this raw material are not satisfied. Unless special measures are taken to restore and maintain such plants, their yield will soon shrink drastically because many species have been depleted by excessive picking and may disappear altogether.

10.3 Need for Ecologization of Industry and Transport

To bridge the wide gap between modern technology and environment protection measures, its all-out ecologization is required. This, however, should not be limited to developing production facilities which would discharge harmless substances into the biosphere. Any substances discharged into the biosphere "in the wrong place, at the wrong time and in the wrong quantities" become biosphere pollutants. Ideal ecologization implies that establishment of an industry which would not disrupt the course of any natural processes and, correspondingly, would not have a negative effect on the vital activity of living organisms. This is possible in two cases: (1) the complete isolation of industry from life's environment, i.e., the development of an isolated technosphere within the biosphere; and (2) the "blending" of industry with the biosphere in the form of certain links in the natural circulation of substances.

There are several stages and ways of transition to such a technosphere. These are primarily waste-free production processes, minimization of the various types of pollution, recultivation of biospheric regions badly affected by industry and reduction of the accidental death of living organisms, caused by production processes.

It is much more difficult, however, to effect industry's "blending" with the biosphere. For the time being, we can only refer to some initial steps in that direction (e.g. De Waal and Van Den Brink 1987).

10.3.1 Towards a Waste-Free Technology

Commoner aptly wrote in his popular book *The Closing Circle* (1971) that the best achievements of our technology — the automobile, jet aeroplane, electric power station, industry as a whole — are our chief failures with reference to the environment. Yet its pollution with harmful substances that we have repeatedly referred to above is not the only ecological danger. Estimates show that the end product consumed by society accounts for as little as 1–1.5% of the total volume of matter involved in industrial production. The rest is production wastes, which sometimes have a disastrous effect on natural ecosystems. Today, any industrial technology must be assessed by the level of its ecological danger, i.e. the quantity and quality of wastes it produces.

Our technological processes are biospherically open. The task is to make them biospherically closed. Ecology-oriented technology must ensure the use of all products obtained at the various stages of production in industrial cycles similar to those observed in natural systems. The wastes of one type of production should serve as raw materials for another. This is the main idea of "waste-free territorial-industrial complex" (Laskorin 1978).

Back in the late 1970's, 83% of water was reused in the chemical industry of the USSR, whereas in its petrochemical and oil-refining industries the share of recycled water was as high as 90–95%.

Transition to complete cycles, a decrease in the resource-to-product ratio and reutilization of resources would all mean transition from extensive to intensive development in the field of technology. Regeneration and reutilization of consumption wastes will complete the technological cycle. The conjugation of many technological cycles and waste-free territorial-industrial complexes within one system will serve as a basis for the development of a technosphere which would also include energetics, transport and, at least partially, agro-industrial associations. Of course, the technosphere will not be entirely isolated from the biosphere. Yet it should be adequately isolated not to disturb the ecological balance in the biosphere and its natural development.

10.3.2 Socio-Ecological Motivation of Production

At the modern stage of the man-biosphere relationship, questions such as "what to produce?", "how to produce?" and "how much to produce?", take on not only socio-economic, but also clearly ecological significance. Society must not produce articles which are "short-lived" or not really necessary.

In the context of the biosphere's limited resources, it becomes relevant to consider the question of whether the production of articles which are not really needed and the development of technologies which are not vitally important are feasible. In the United States, for example, only one out of 40 articles developed for production reaches the market, while from 40 to 80% of consumer goods and from 20 to 40% of manufactures that have reached the market fail to find a buyer.

This is the price (not only economic but also ecological) the capitalist society pays for the excessive variety of goods.

Any big production facility may serve as an example of the production of things which are not really necessary. The production of household plastic articles from the chemical industry's waste products is becoming increasingly popular in all developed countries. When only wastes are used for the purpose, this can be justified somehow. But the production of plastic articles, such as toy wrapping and small household articles, is increasingly becoming an end in itself, in which case it turns into an undesirable, rather than useful, undertaking. Indeed, on the one hand, non-renewable natural resources are squandered to make throw-away articles, on the other, an extremely stable plastics pollution of the environment occurs. Many similar examples, the overproduction of military hardware and equipment being the most important one, point to the need to develop production on the basis of man's real requirements, on the principles of universal security, social equality and duly accounting for the biosphere's limited ecological possibilities.

10.3.3 Industry's "Blending" with the Biosphere

To produce food protein, it is possible (as is the case today) to organize the building of expensive and resource-intensive expeditionary floating bases and dispatch them thousands of miles away from the home country to catch, by employing advanced (and also expensive and resource-consuming) techniques, fish and invertebrates in the World Ocean. (Let us remind the reader that one third of the marine products thus procured is a source of the food protein used currently for fodder). It is also possible, while reducing many times the expenditure of metal and other basic materials, to organize microbiological production of the same fodder protein by using the chemical industry's waste products, low-molecular alcohols, natural methane and even hydrogen. The latter way is surely much more in keeping with the need to ecologize industry and man's economic activities on Earth in general.

Here are two more major ways of ecologizing industry, which are also connected with technical microbiology.

It is common knowledge that large-scale open-pit mining does great harm to nature. It involves changes in the land hydrological regime over vast territories, the scattering of millions of tons of substances, the exclusion of huge areas of arable land from circulation, to say nothing of the liquidation of habitats (sometimes unique) for animals and plants. At the same time, however, various techniques exist for extracting metals, such as copper, zink, ferrum, aluminium, tin and even gold, by using biometallurgy, i.e., microbiological leaching with no stripping and mining.

It is also common knowledge that the chemicalization of agriculture is one of the chief sources of environmental pollution. The use of chemical fertilizers, including nitrogenous substances, is most widespread. Hundreds of millions of tons of nitrogen fertilizers are scattered annually over the Earth's fields, which are mostly not assimilated by plants and washed out into reservoirs, disrupting the course of natural processes. In the USSR alone, several billion roubles is spent to this end. Yet it is known that nitrogen-fixing bacteria that live in the soil fix the chemical in no less a quantity than that introduced by man (in any case the figures are comparable). Theoretically, it is possible to activate the work of these bacteria, which is incomparably safer ecologically than the production of nitrogenous fertilizers. In the global economy, factory-made nitrogenous fertilizers will largely be replaced in future by microbiological techniques.

The production of many plastics and wrapping materials requires rapid and effective ecologization. The main demand is that they should decompose rather quickly in natural conditions with no release of components harmful to the biosphere.

It is not fortuitous, it seems, that the above examples of "blending" production processes with the circulation of substances in the biosphere were mostly centred around the microbiological industry. It is quite possible that it is microbiology and biotechnology that will substitute ecologically acceptable industrial processes for the current anti-ecological ones.

10.3.4 Roadsides as Possible Living Nature Reserves

The rapid development of transport and its ever-increasing influence on living nature and the biosphere as a whole (see Section 4.2.5.2) compel us to think about the need to ecologize it, too. So far, our attention has been mainly concentrated on the development of ecologically "clean" transport facilities which would pollute the air minimally. But other problems also arise here in connection with the protection of the environment. In industrialized European countries, the total road area amounts to 2–3% of the land. In the United States, the figure was nearly 10% in 1980, excluding Hawaii and Alaska. These territories offer some prospects for the preservation of certain amount of species.

Roads in any country include not only traffic lanes, but also roadsides, ditches and, as a rule, green zones, i.e. shrub- and tree-covered sidings. In a number of cases roadside plantations may serve as habitats for some wildlife species.

Of great importance are green roadside strips in big cities and urban agglomerates stretching over tens and hundreds of square kilometers. They may serve not only as habitats but also as the only flight paths for small birds, as well as the routes of migration and dispersion of many other animal and plant species (Anderson 1977).

Also important are roadside tree plantations in steppe and semi-steppe regions. In the setting of intensive agricultural production, they may serve as reserves for wild vegetation and the last shelter for many animal species forced out from big agrocenosis spaces.

Making the fauna of roadside strips richer has also "the other side of the coin": an increased number of accidental killings of animals by motor vehicles and other transport facilities. In a number of countries, special traffic signs on highways warn the drivers that large animals, such as deer, elks, wild boars and roes, may cross the road. Low fences are put up along highways to prevent smaller animals from being crushed by passing cars. In the FRG, more than 1000 km of motorways had fencing (including electric) in 1980 to ward off rabbits, hares and roes. Car headlights reflected by small roadside reflectors (preferably red) are very effective in scaring animals away from passing cars. Also effective are miniature ultrasound emitters installed in cars, which generate 17,500 Hz sound at speeds exceeding 50 km h^{-1}. All this is economically sound, because on the roads of North Rhine-Westphalia (FRG) alone losses due to animal fatalities on the road (nearly 100 daily) were estimated at DM 10 million (Heinemann 1981).

In areas of mass animal migration it is necessary to build special passes, as has been done in Alaska and Taimir to enable reindeer to cross the pipelines, or in Holland, Australia, the FRG, France and some other countries, to make it possible for toads, frogs, tritons, tortoises, opossums, hedgehogs and other animals to cross motorways. However, many problems still exist.

10.4 Need for Ecologization of the Economy

Economic stimuli have always been among those most effective for the development of any society. This necessitates their optimal use in solving the urgent problems of the protection and rational utilization of the animal and vegetable kingdoms and also of their habitats (e.g. Fischer, Peterson 1976). The preservation of the variety of animal and plant species is the main criterion for the quality of the living nature environment. This is so because of the specifics of biological resources: a lost species cannot be restored. Another criterion for the quality of the environment, no less important and closely connected with the first one, is the preservation of the ecological balance in a certain area or region, or in the biosphere at large.

Let us consider some aspects of the application of the above principles to the practical ecologization of the economy.

10.4.1 Determination of Relative Values in the Protection of Living Nature

At least three criteria to be observed by society as regards individual species can be formulated: (1) each species must be preserved (excluding those dying out naturally); (2) the intraspecies genetic diversity must not be lost (id est an adequate number of variable populations must be preserved); (3) outlays for the exploitation of any population of organisms (hunting, fishing, recreation or any other method of nature utilization) must include funds for reproducing populations and maintaining their stable level.

If these criteria are taken to determine the relative value of protecting certain species, various approaches are possible. First of all, priorities may be based on the economic importance of species: those having a higher economic value have an edge over the other species.

The magnitude of possible genetic loss due to the disapperance of a particular form is a more accurate criterion for determining the value of this disappearing species. It is clear that when determining socio-economic values or the priority of protection expenses, a species representing a disappearing order has an advantage over a species which is the only representative of a disappearing genus. A species representing a disappearing genus has an advantage over a species representing another genus where the number of species is bigger. Finally, a disappearing species must be saved more actively than a disappearing subspecies.

With reference to genetic diversity protection as a chief task in the protection of living nature, we come up against the need to make an economic assessment of social values and benefits. Mankind needs the genetic diversity as a basis for creating real economic values. Yet as a particular species becomes increasingly rare, its ability to be of real use for society diminishes. Species on the brink of extinction and therefore excluded completely from the economy are more likely today to have a negative economic value: society has to spend considerable sums to protect them. But their relative social and biological value increases in proportion to their rarity.

The character of a species' distribution should also be taken into account: it should be established whether the given species is in trouble only on the territory of a particular country or even a particular region within its borders. The threat of extinction facing endemics is many times more alarming than the loss of species whose extinction in one particular country will not result in their loss for mankind at large. There can at least be a two-pronged approach to determining priorities depending on the geographical level of the protection status: one is based on the primacy of the International Red Book (the global level of protection), the other on the dominance of the priority protection of forms whose magnitude has shrunk dangerously in a particular region (the local level of protection). For example, the brown bear (*Ursus arctos*) is a game species on Kamchatka, whereas in the Baltic Republics it is absolutely protected. Different levels do not contradict, but rather supplement one another.

The relative value of a species will not be fully established if we do not line up the representatives of all the major groups of the organic world in accordance with their representation in the biosphere. If the total number of insect species is at least three to four million, of higher plants some 300,000 and of vertebrates nearly 5000, the relative evolutional-biogeocenotic value of a vertebrate species might be higher than that of higher plants or insects.

Besides the above briefly outlined approaches to determining the relative value of a species and, correspondingly, establishing priorities in taking protection measures, a comprehensive evaluation of living forms should also take other criteria into account.

Specifically, account should be made of the main causes of reduction in the population of a particular species. For example, extinction may threaten a species as a result of the destruction of its habitat, the adverse effect of intruding species or the loss of food sources (see Chap. 4.2), i.e. causes which cannot be eliminated immediately. In such cases measures to protect these species should be more extensive and costly than, say, those taken to save species whose population is shrinking as a result of intentional or accidental actions by man (which, as a rule, may be halted relatively easily and swiftly).

However accurately the relative value of species and, correspondingly, the priority of protection measures might be established, in reality some of them cannot be saved even by taking most costly measures. In this case it seems to be more feasible instead to spend money on protecting those species which are more likely to be saved. On the other hand, no one would risk arguing against the US attempt (with no guarantee of success) to save the California condor (*Gymnogyps californianus*), threatened with total extinction (see Section 4.1.6 in Chapter 4), even at the cost of a 25 million-dollar project (i.e. more than US $1.1 million for each living bird). Here is another example showing that final assessments are hard to make. We all consider the success of Project Tiger in India as a major national achievement. Thanks to effective measures taken by the Indian Government, the population of tigers has increased from 1800 to 4000 species, although the "cost" of each tiger amounted to nearly US $20,000 (the average daily earning of Indians does not exceed one dollar).

10.4.2 Determination of the Absolute Value of a Species

At least one interesting attempt was made to determine the absolute value of species (Reymers and Shtilmark 1978). The line of reasoning is as follows. On the basis of Rule 10, which has repeatedly been verified empirically, a 10–20% loss of the existing species would disrupt the ecological balance on our planet. Twenty percent of the nearly 1.5 million species described so far makes 300,000. Their disappearance is equivalent to a loss of the world national product as a result of a total destruction of the Earth's biosphere. According to UN figures, the world national product in the 1980's was estimated at US $3400 billion. Hence the average conventional cost of a species is $11.3 million:

$$\text{Species Value} = \frac{\text{World National Product}}{\text{Species Critical Number}}$$

From five to ten million species are probably not known to science. This adjustment made, the value of a species calculated by using the above technique may range from $2.6 million to $1.5 million.

The above evaluation is, of course, very approximate and speculative. Moreover, account should be made of the fact that the cost of the world national product is growing rapidly. Besides, it is more feasible to make calculations on the basis of the world national wealth, which is many times more than the world national product. This leads to the conclusion that the actual value of a species has to be many times higher than given above. This approach to evaluating species and also ecosystems is applicable to any individual regions.

Approximate and relative as it might be, this calculation is indicative and gives certain basic data to "justify" society's expenses for preserving the disappearing species of animals, plants, fungi and prokaryotes.

10.4.3 Ecologization of Industrial Product Cost Estimates

Nature, living nature included, is the prime source of the implements and objects of labour. If natural resources were unlimited as compared with the volume of society's needs, they could remain free. If, however, they are scarce or in short supply, they inevitably have to be paid for. This is a matter of continuous discussions among economists, both in socialist and capitalist countries, which fact indicates that this economic problem of nature utilization is global in character.

Today's situation is such that the problem of paying for air will probably become a grim reality for the living generations (Fedorenko and Reymers 1981), rather than some fictional story told in the 19th century.

The problem of paid natural resources has several interim solutions apparently making it possible to combine the interests of preserving the system that ensures society's viability, and the interests of continuous technological progress. One such solution would be based on collecting payments only for those natural resources whose preservation and reproduction entails certain expenses on the part of society. If a State protects and preserves its forests to maintain the global oxygen balance, it is entitled to a certain economic compensation to be paid by other states interested in air oxygen. If one region consumes oxygen produced by forest plantations in the neighbouring regions (especially northern ones), this means that there are grounds for establishing specific economic parameters for inter-regional ecological contacts within one country.

The cost of the final product in any enterprise (and not only that of the direct use of living resources, such as felling, hunting or fishing) should also include "nature-intensity" indicators and the ecological cost of production. Society is not indifferent as to how much oxygen or water has been consumed to make a particular article, and what indirect consequences this will lead to as regards the animals and plants in the region, country or the world at large.

"Nature intensity" should also include the "waste intensity" of production, i.e. the quantity of pollutants released as a result of the production and exploitation of a particular product, be it a jet-liner or a piece of man-made leather. A piece of natural leather may turn out to be more profitable socially, ecologically and economically than a piece of man-made leather of the same size, because the polymer production has caused a massive discharge of harmful wastes into the environment. The cost of off-shore oil, for instance, should include, in whatever form, the value of the thousands of birds that perish in waste gas torches or oil spills.

The production of synthetic rubber is considered more profitable today than the cultivation of hevea rubber trees and the production of natural rubber. The economic correlation of these two production methods, with "waste intensity" duly accounted for, is as yet unknown. Large livestock farm complexes are believed to be more profitable for meat production. Yet, if their "waste intensity" and "ecological cost" are taken into account, it will become clear that relevant indicators vary depending on many factors. Under certain conditions, these factors alone may make it unprofitable to expand the farm or the complex further and concentrate the wastes in one place.

The list of similar examples may be expanded. All of them show that the contemporary system of economic indicators does not explicitly provide for this important aspect of the utilization of natural resources: the disposal of production wastes which are not the end product of a particular enterprise and cannot be stored on its territory for a sufficiently long time. Hence the conclusion that an indicator should be worked out and used, which would include ecological damage in the product unit cost.

Economists should determine whether the expenses related to the direct or indirect use of living natural resources, such as a reduction in the population of certain species, a decrease in the amount of oxygen produced by plants or a shrinkage of the recreational capacity of a locality, should be included in the product cost or paid for out of society's centralized revenues. Both options are possible, as well as some interim solutions. The first option, however, may prove to be most effective from the viewpoint of its usefulness for living nature because it could quickly result in a reduced demand for expensive (harmful to nature) products. As a result, the producers' efforts would be directed towards developing less harmful products, rather than expanding the production of nature-hazardous articles (Lemeshev 1978).

The protection of nature cannot be unprofitable for a society or humanity at large. What is important is to make this general proposition a real economic lever to protect living nature.

10.5 Need for Ecologization of the Urban Environment

The ecologization of the urban environment is an important aspect of the protection of living nature. A considerable part of the urban area is taken up by public gardens, parks, lawns, ponds and other such places, plants and animals being their inalienable elements.

In keeping with the standards now in force in the Soviet Union (1978), the per capita area of green plantations in the mid-country should be some 20 m² in towns and cities with a population of up to one million. In 1985, this area in Moscow was nearly 25 m² or some 40 m², forest-parks within the city's boundaries included. Town ecosystems depend entirely on Man (watering, periodic planting, garbage collection, etc.). But Man, too, depends on them. As a being that has emerged during the course of biological evolution, Man finds it very difficult to live in the town without at least brief contacts with Nature. This alone gives grounds to assert that the small islands of living nature in towns and cities are absolutely necessary. As the green zone area there increases, the variety of urban fauna may also increase. In Soviet towns with a medium level of population density, for example, up to one third of the birds living in the given locality find their habitats there. Swifts, swallows, tomtits and other insectivorous birds have for centuries been permanent residents of cities, towns, villages and other human settlements in many countries (up to the moment chemical pesticides were invented). The total number of nesting bird species in some large Soviet cities exceeds 150–170 (Table 10.2).

The summer-time ornithofauna of a European city normally includes up to one third of the birds of the given geographical zone. In winter, however, the number of species in some towns is not less but more than that in their natural habitats. In Saratov, for example, 42 species have been seen in winter, in Volgogard 41, at Akademgorodok near Novosibirsk 48, and in Omsk 29. This is almost two times more than the number of species hibernating there in the natural environment.

In Leningrad 10 bird species have disappeared over 100 years of observation, but 18 new species have appeared. In Moscow, 16 new species have also appeared in the period between 1970 and 1980, including wood pigeons, collared turtledoves, golden-eyes and blackbirds (Blagosklonov 1980). This gives hope that man and wildlife may co-exist; unfortunately this possibility is now not realized (see Section 4.1.6).

Table 10.2. Numerical composition of ornithofauna in some major cities by the late 1970's. (Based on figures from various sources)

City	Total number of species	Including	
		Nesting	Staying temporarily
Frunze	?	129	?
Moscow	177	110	67
Omsk	169	83	86
Novosibirsk	158	22	136
Lodz (Poland)	158	108	50
Tomsk	135	77	58
Leningrad	138	73	65
Saratov	133	66	67
Volgograd	87	55	32
Brno (Czechoslovakia)	77	56	21
Kishinev	72	20	52
Sverdlovsk	70	37	33

People in many countries have for many years been successfully attracting insectivorous birds to nest in city parks. Many birds, such as several wild duck species, have been successfully introduced into the town fauna. Less known are methods which do not imply special introduction. Urban ornis may be enriched by merely creating conditions favourable for nesting. Swans (*Cygnus olor*) are now increasingly frequently making stopovers or even nests in town ponds in the Baltic Republics (Elgava) and near the White Sea (Severodvinsk). Magpies (*Pica pica*) and jays (*Garrulus glandarius*) arrive and nest more frequently now in towns in the European part of the USSR, in the Urals and West Siberia, while azure-winged magpies (*Cyanopicus cyanus*) have become more frequent visitors to towns in the Amur region (Priamur'e). Common eiders (*Somateria molissima*) have for a century been a nesting species in many settlements in Northern Europe. Various wild ducks and gulls, and also several goose species are now a common element of the urban ornis of many European towns. In the mid-1970's, the Leningrad and Novgorod regions of the USSR were populated by white storks (*Ciconia ciconia*) thanks to the provision of artificial nesting places for them: more than 95% of the known nests there were built by birds on waggon wheels raised on poles in villages and settlements (A.S. Malchevsky, personal communication).

Analysis has shown that the largest colonies of black-headed gulls (*Larus ridibundus*) in Sweden are located near big cities, rather than far away. Over the past 50 years, the total number of such colonies has increased from 50 to 800. It should be noted that Europe's largest colony of black-headed gulls (up to 15,000 nesting couples in 1980) is only 15–17 km from the boundaries of Moscow, in the territory of the town of Lobnya, where there live as many gulls as townsfolk. The count of gulls living in towns in Britain and Ireland has shown that the number of herring gulls (*Larus argentatus*) and lesser black-backed gulls (*Larus fuscus*) nesting on buildings is increasing every year.

All these and many other similar facts show that two opposite processes are under way simultaneously: ornithofauna is becoming scantier as a result of urbanization and the ensuing recreational load on public parks, on the one hand, and birds are "growing into" the fauna of towns and cities, on the other. It seems that a moderate level of urbanization which results in a variety of habitats (something like the ecotone effect in ecology) is conducive to a greater variety of ornis (Fig. 10.3).

Some increase in the variety of urban fauna is also characteristic of mammals. Many animals successfully adapt themselves to living in close proximity to Man (naturally, provided people do not exterminate them). Squirrels live in the parks of Alma-Ata, Moscow, Washington and many other big cities throughout the world. People know less about badgers (*Meles meles*) living in the basements of apartment houses (in Copenhagen alone more than 100 families of badgers are living), otters (*Lutra lutra*) and beavers (*Castor fiber*) living in large human settlements, martens and foxes living in big cities "in man's shadow." In 1977–1979, for example, the number of common foxes (*Vulpes vulpes*) in Greater London exceeded 3000. They ousted some of the stray dogs and cats from the city

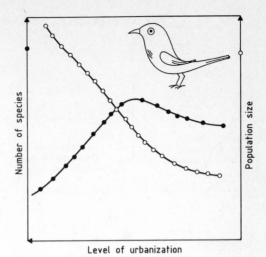

Fig. 10.3. Correlation between total bird populations and species variety at different urbanization levels. (Yablokov and Ostroumov 1983)

fauna. In Melbourne, duckbills, echidnae, koalas and some small kangaroos live in freedom. Thousands of foxes and skunks live in major cities, such as Los Angeles and San Diego; hundreds of rock martens (*Martes foina*) live in Voronezh, Kishinev, Odessa, Brno and some other European cities.

It is safe to say that we are witnessing the urbanization and, to some extent, synathropization of rock marten, common squirrel, black-footed ferret and Russian polecat on the Russian Plain, grey squirrel, badger, common fox and some other mammals (including several chiropterous species) in Western Europe. In Japanese cities, the process involves racoon dogs, white-cheeked flying squirrels and chipmunks; in Indian cities, two macaque species (*Macaca mulatta* and *Macaca radiata*), whose urban populations have for many centuries differed markedly from the wild-life species.

The facts show that there are five sources for the formation of an urban natural environment (Fig. 10.4).

The urbanization of species is based on the natural selection of individuals not afraid of man and on the possibility of using new sources of food and shelter. Animals which are panic-stricken by city noise, moving transport or man's presence near their nests or holes cannot live in town.

The selection of animals individually capable of living in town is only the first stage of the urbanization of fauna. The second stage is the creation of stable and numerous reproducing urban populations. The third stage is the formation of a real town ecosystems with all its components (producers, consumers, reducers) and characteristic features (stability, plasticity, etc.). It differs from natural ecosystems in that man's direct participation is an important (at times decisive) factor at various "entries" to and "exits" from this system. A part of 600–900 g of wastes and garbage produced daily by each city dweller in developed countries plus organized feeding may play the role of producers here.

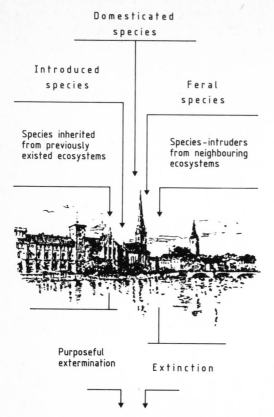

Fig. 10.4. Chief sources of the formation of the species composition of urbanized territories. (Yablokov and Ostroumov 1983)

The city dweller's need to regularly "communicate" with nature seems to be one of the most vital needs, yet not fully comprehended so far. It is quite possible that the lack of such communication is the root of the increased number of behavioural disturbances and the growing spread of certain "urban" diseases. Forest recreational zones are needed not only in the suburbs, but also near each town district. In the final analysis, this would mean limitation of the general density of population in towns when their housing and development programmes are devised.

Urban ecologization should also imply care for the preservation and consolidation of ecosystems ensuring the purity of natural water and recycling wastes, drainage water, etc. The first group of ecosystems traditionally attracts closer attention, but here, too, there is room for improvement (most probably through imposing a stricter protection regime, paying special attention to eutrophication control and the prevention of water reservoirs from being polluted with industrial and agricultural wastes). The second group of ecosystems performing purification, detoxication and other such functions has so far been studied to a lesser degree and, generally, is given less attention. It seems that further work in this field must include, among other important measures, analysis of the structure and functions of those blocks of purifying ecosystems which transform or absorb pollutants. Such blocks include soils, silts, destructor- and reducer-organisms, including fungi, unicellular water-plants and prokaryotes. The pollutant-absorbing and pollutant-degrading capacity of biogeocenoses will have to be evaluated and enhanced.

To sum up, the necessary directions of work on urbanized territories include: purposeful enrichment of their fauna and flora; formation and maintenance of the biocenoses of urbanized ecosystems; ecologization of city-dweller's consciousness; preservation of natural ("pre-urban") biogeocenoses within city limits; ecologization of town planning, housing and communal services in general.

10.6 Need for Ecologization of Medicine

One may say that the protection of living nature in its entirety is, in a sense, medicine-oriented: according to the WHO's definition, health is a state of complete physical, spiritual and social well-being (Health 1972). It is clear, however, that man cannot attain either social or spiritual well-being without preserving living nature. Moreover, the physical well-being of both individual people and their populations is closely connected with the preservation of nature in all of its variety. People's health is a good indicator of the ecological well-being of the environment.

Many environmental factors have a direct and negative effect on the organism of man as a biological being (see Chaps. 2, 3). Many dangers facing living nature may be studied on the sad experience of humans who are affected by the very same pollutants (mutagens, carcinogens and allergens) which affect other living creatures in principally the same way.

Man is a kind of "walking biocenosis": millions of microorganisms live on the surface of his body and in its cavities (intestine, lungs, throat, nasal meatus), most of them being his symbionts. They have adapted themselves to living on Man and seem to largely determine his physical well-being (suffice it to remind the reader of the immense and vitally important role of the intestine microflora for normal digestion).

Although we know extremely little of the delicate links between Man and numerous unicellular and multicellular species (the latter category including microscopic ticks and helminths), a general evolutionary-ecological approach prompts the conclusion that this group of organisms is closely interdependent. The presence of Man's own bacteria on the surface of his skin and his mucous membranes is a solid biological barrier preventing alien and health-hazardous creatures from penetrating into Man's organism (Marples 1965; Maibach 1965).

During the course of one experiment, a concentrate of the typhoid bacterium (*Salmonella typhi*) was put on the surface of human skin and on a glass plate. In

20 min, the overwhelming part of the bacteria on the skin perished, whereas most of those on the plate lived (Andrews 1977). Many bacteria living on human skin produce specific antibiotics protecting wounds from pathogenic bacteria. This is why it often happens that by taking excessive doses of antibiotics and thus killing our own bacteria, we reduce our own resistibility to various diseases.

The whole conception of parasitism will have to be reconsidered in the near future. According to this conception, the liquidation of "parasites" is always and absolutely useful. Yet, in view of the fact that the master's death is usually not to the parasite's advantage, a selection vector is formed to develop such parasite properties which would make their presence somewhat useful for the master or would drastically reduce their virulence (for a survey see Kennedy 1978). This was the case with the rabbit myxomatosis in Australia. This was, evidently, also the case with classical Asian cholera, which killed millions of people between 1817 and 1926 and was then "replaced", before our very eyes, by a less virulent cholera — El Tor (Baroyan 1982).

What has been said above indicates the need for an increasingly ecological, ecosystems approach to many medical problems. Some of them are directly connected with the protection of living nature. Above all, they include the problem of parasite devastation and infection eradication. One can only agree with Baroyan and Lepikhov (1975), who emphasized the ecological danger of breaking the links between man and his immediate living environment. The only way out is to reduce mass diseases to a level that would remove the danger of epidemics. An ever-increasing number of microorganisms are now in this situation: the pathogens of diphtheria, poliomyelitis, pertussis, tetanus, plague, pox, cholera, etc. The complete liquidation of smallpox is usually regarded as a major success of modern medicine (the last epidemic was recorded in 1977 in Somalia). But 2 years later in Nigeria, a case of monkey pox was registered, its victim being a man. Pox-like viruses are widespread among animals, including cows, camels and rodents. So, it is quite possible that in a situation where there is no immunity to the now non-existent smallpox, favourable conditions will emerge, facilitating the virus strains which are so far not dangerous to develop into strains which will be even more dangerous for man than the smallpox that has been eradicated.

The future world will be a world of controlled, but not eradicated infections. This pertains even to rabies, whose pathogen is always present in living nature. Carnivorous animals, mainly dogs and allies, are rabies' carriers and "custodians." Attempts were often made up to the mid-20th century to control rabies by means of complete extermination of predatory animals. One of the latest attempts was made in 1952. In a bid to liquidate an outbreak of rabies in the Province of Alberta, Canadian authorities began shooting beasts of prey en masse. They exterminated some 50,000 foxes, 100,000 coyotes, 4,300 wolves, 7,500 lynxes, 1,850 bears, 500 skunks, 64 cougars, four badgers and one glutton. Yet these measures only resulted in the eradication of rabies in few regions. On the other hand, mouse rodents and also deer and elks had multiplied enormously as a result of the massive extermination of predatory animals. The propagation of wild hoofed animals led to a depression of grass and shrub vegetation and great damage to forests, which in turn brought about a catastrophic decrease in the population of wild hoofed animals, which lasted for many years. The propagation of rodents resulted in the massive reproduction of foxes (those species which had learnt how to avoid poisoned baits). The seat of rabies had rapidly re-emerged (MacDonald 1980).

The right solution appears to have been found by Polish medical authorities and zoologists, following the outbreak of rabies in Poland in the 1960's. They completely exterminated only the rabies-carrying foxes and all stray dogs and cats. The rabies pathogen that had remained in nature circulated exclusively among the fox population. As a result of natural selection, they developed resistance to it, and so the epidemic was stopped with no harmful ecological consequences.

It is known that the full extermination of ticks by using DDT in the area of a tick-borne encephalitis outbreak does not check the virus circulation by blood-sucking dipterans, tsetse flies and many other species which are the objects of repression in medicine. This also holds true of the pathogen of plague. All this shows that the medical zoologists' hopes to check the spread of a pathogen by fully exterminating some particular virus carriers are not justified from the ecosystems point of view: viruses easily find alternative ways of circulation, whereas an all-out battle against virus carriers does not yield the desired effect, causing substantial damage to the environment. Moreover, an ever-increasing number of insects successfully develop a resistance capability vis-à-vis the increasingly more potent insecticides used against them (Fig. 10.5).

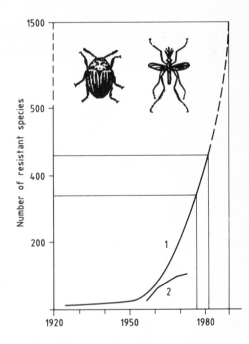

Fig. 10.5. Increase in the number of insecticide-resistant insect species from 1920 to 1985. *1* insects in agrocenoses; *2* infectious disease carriers and house dwellers. (Based on data from various sources in Yablokov and Ostroumov 1985)

The future replacement of all-purpose antibiotics (which prove to be dangerous in principle because pathogenous microorganisms easily develop resistance to them) by the biochemical products of animal and vegetable origin is an important aspect of the ecologization of medicine. Some shifts in that direction have already been made. Half of the medical preparations produced in the world today are based on products obtained from various living organisms. As has been mentioned earlier, botanists and medical workers know of some 2500 species of the Soviet Union's flora possessing medicinal properties. Of these, not more than 200 species are used so far to produce various medical preparations, including some 50 species cultivated on special farms. It does not appear fantastic that in the foreseeable future the number of plant species used in Pharmacopeia and cultivated as crops (and thus saved from extinction) may increase by an order.

The wide use of antibiotics in food production (as additives to cattle fodder and bird feed and conservants) involves another important ecological problem of medicine. By 1984, experts in the United States had established, finally and conclusively, a connection between the use of antibiotics in cattle breeding and poultry farming and the development of resistance-capable forms of pathogenic microorganisms. As a result, people taking the same or similar antibiotics (usually penicillin, tetracycline and their derivatives) do not receive the expected positive effect. Moreover, they risk making their organism the most suitable medium for the unconscious selection of strains of microorganisms, such as *Salmonella* and *Campylobacter*.

The ecologization of medicine also involves the benevolent effect of the environment on the rapid recuperation of patients suffering from serious illnesses. It has been established statistically that post-infarct patients, with small domesticated rodents, dogs, cats, birds or even iguanas living at their homes, have better chances for swift and successful recuperation.

10.7 Need for Ecologization of Politics and Law

The global nature of environmental protection problems is rapidly moving them to the foreground as the most urgent political issues. We are witnessing a phenomenon which can be called the "ecologization" of politics and law. But it is confronted by many difficulties and obstacles, some of which can be overcome comparatively easily. For example, each country having animals migrating across its national border may conclude special agreements with all the other countries where these animals stay at the various periods of their life cycles (hibernation, reproduction, migration, etc.). To this end, relevant provisions of international law have been adopted: Principle 21 (The Principle of Good-Neighbourliness) of the Stockholm Declaration on the Human Environment (1972) says that each state shall undertake to ensure that the activity within its national jurisdiction or control should not be detrimental to the environment of other states or regions outside the limits of the national jurisdiction.

The adoption of a system of comprehensive international agreements on the protection and utilization of both the animal and vegetable kingdoms under UN auspices, on the one hand, and the development of bilateral agreements between countries with a common fauna, on the other, will ensure a dependable international legal framework for the protection of living nature.

In our opinion, principal difficulties in this field stem from the need to change certain traditional approaches to the utilization of nature, now global in character.

The first difficulty consists in developing political conceptions which would allow for the "principle of compensation for the preservation of resources". It is known that many smaller countries (especially in the tropical and subtropical zones) possess a unique vegetable and animal world. To preserve this wealth, economic activities should be drastically limited and a considerable part of the territory should be excluded from such activities to set up a network of protected territories. It is expedient for the world community at large to encourage such countries to preserve this invaluable evolutionary wealth as much as possible (this wealth being, in fact, part of the world community's living natural heritage) and to somehow compensate such countries for certain restrictions imposed on the development of their productive forces. This, of course, does not mean that their social and economic development should be halted, but the direction of such development should be in keeping with the interests of mankind at large, rather than contradict them, as was the case with Madagascar (as a result of the extensive development of agriculture, most forests were felled and the island's unique biota has become catastrophically poor).

Another difficulty also calling for a certain revision of the existing political conceptions involves the development of the "principle of the common world heritage of living nature." It implies recognition of the right of any country to benefits from the exploitation of the World Ocean's natural resources beyond States' national jurisdiction (see also Chap. 7).

The compensation principle will also have to be extended to the mechanisms of maintaining the quality of all the basic elements of the natural environment: the sources of free oxygen, freshwater, climate-forming territories, etc. All this is not such a far-away prospect as it might seem on the face of it. It is known that the United States suffers from the shortage of oxygen produced on its territory and it "inhales" some 40% of oxygen generated by the World Ocean, Canadian, Siberian and Amazonian forests. This determines the USA's direct interest in preserving a considerable part of the forests in all the above regions and in maintaining the purity of the World Ocean as conditions for the country's further economic development.

It is also known that in the mid-1970's a wave of farmer protests swept Mexico: Mexican farmers were angered by the large-scale condensation of moisture to water in the arid lands of the US southern states. They believed (most probably, with good reason) that this artificial overhead irrigation in the United

States would inevitably dry up the atmosphere over Mexican territory even more. Similar conflict situations in the utilization of natural resources will be increasingly repeated in future, simultaneously with man's growing ability to affect natural processes in the Earth's biosphere.

The development of the international connections at large may entail immense and important efforts to unify national legislations pertaining to the protection of the animal and vegetable kingdoms. This unification would largely enhance the effectiveness of a system of measures to protect living nature on both regional and global levels. A good start was made by the inclusion in the Final Act of the Conference on Security and Cooperation in Europe (1975) of a provision spelling out the need to draw closer together the legislations of individual countries dealing with environment protection measures.

Besides the above conceptions, life will demand the adoption of quite new norms of environmental international law. They would include a provision concerning a State's "maximum (ultimate) responsibility" for preserving species whose natural habitat is exclusively on its territory. In 1980, the IUCN Commission on Rare Species proposed that the practical elaboration of this legal conception should be started by declaring the States' "maximum responsibility" as regards the following 12 species:

1. Leadbeater's possum (*Gymnobelideus leadbeateri*) Australia
2. Cuban solenodon (*Solenodon cubanus*) Cuba
3. Golden tamarin (*Leontopithecus rosalia*) Brazil
4. Javan rhinoceros (*Rhinocerus sondaicus*) Indonesia
5. Indian dolphin (*Platanista indi*) Pakistan
6. Giant golden mole (*Chrysospalax trevelyani*) South Africa
7. Aye-aye (*Daubentonia madagascariensis*) Madagascar
8. Lion-tailed macaque (*Macaca silenus*) India
9. Simian jackal (*Canis simensis*) Ethiopia
10. Iriomote cat (*Felis iriomotensis*) Japan
11. Arabian tahr (*Hemitragus jayakari*) Oman
12. Russian desman (*Desmana moschata*) USSR

The level of legislative, political and public action to protect the environment is increasing enormously throughout the world. Public movements, such as the Greens, have taken up firm positions on the political scene of many West-European countries, and ecological problems are now prominent in the political programmes of all major national parties. In countries with centrally-planned economies, environmental protection has become one of items in the national plans for economic and social development.

Finding solutions to environmental problems in the politically uneasy world torn apart by economic and social contradictions is a major factor that should rally all countries.

11 Ways of Preserving Living Nature

However dangerous or even tragic anthropogenic phenomena might be for living nature, Man is able to restore the population of species which are on the brink of extinction. The examples are many, exceeding by far those usually referred to by specialists. Some are included in Table 11.1, which convincingly shows that the population of species is restored if Man takes special and active measures. This process often occurs in countries with the highest population density. Over the second half of the 20th century, in Britain, for example, there has been a steady growth of the population of some species, such as the wild cat (*Felis silvestris*), the polecat (*Mustela putorius*) and the marten (*Martes martes*). In 1984, the United States excluded the brown pelican (*Pelicanus occidentalis*) and the American alligator (*Alligator mississipiensis*) from the list of rare species thanks to successful measures to restore their population. There are examples, albeit rare, of the restoration of the population of a species whose numbers have dropped under the pressure of excessive hunting or fishing, without banning such activities altogether (though strict regulations measures were, of course, adopted). This is the case with the Pacific herring (*Clupea harengus pallasi*) in the north-western part of the Sea of Okhotsk, the king crab (*Paralithodes camtschatica*), the pollock (*Theragra chalcogramma*), some seals and a number of other species (Table 11.1). As a result of restoration, ten species and subspecies of mammals and birds have been transferred from the red to the green pages of the International Red Book. They include the Leadbeater's possum (*Gymnobelideus leadbeateri*), the hairy-nosed wombat (*Lasiorhunus latifrons*), the Guadaloupe fur seal (*Arctocephalus townsendi*), two subspecies of deer, the white-tailed gnu (*Connochaetes gnou*), the European bison (*Bison bonasus*), the grass parakeet (*Neophema pulhella*), the Australian flycatcher (*Sophodes nigroullaris*), and the New Zealand starling (*Creation corruncullatus*).

All these and many similar facts make one feel confident that vigorous action to restore the population of rare species is feasible and effective.

Table 11.1. Examples of successful restoration of the population of certain mammals and birds. (After Yablokov and Ostroumov 1983)

Species	Population			Main restoration factors
	Initial[a]	Minimal	Current	
European beaver (USSR)	Hundreds of thousands?	Some 700 (1920's)	Some 200,000	Ban on hunting, organization of preserves, reacclimatization, resettlement
Asian sea otter (USSR)	Not more than 15,000	Some 900 (1950's)	Some 7000	Ban on hunting, organization of preserves
Sable (USSR)	Hundreds of thousands	7000–8000 (1930's)	Some 700,000	Ban on hunting, resettlement, preserves
Amur tiger (USSR)	Several hundred	30 (1935)	Some 200	Ban on hunting, preserves
Bengal tiger (India, Bangladesh)	Some 40,000	1827 (1972)	More than 4000	Ban on hunting, preserves
Far Eastern walrus (USSR, USA)	Some 200,000	Not more than 30,000 (1950's)	Some 200,000	Coordinated national regulation of hunting
Northern elephant seals (USA, Mexico)	Hundreds of thousands	20 (1890)	More than 100,000	Ban on hunting, protection of reproduction sites
Guadaloupe fur seal (Mexico, USA)	Tens of thousands	60 (1928)	Some 2000	Ban on hunting, preserves
Onager (USSR)	Tens of thousands	Some 200 (1950's)	More than 2000	Preserves, resettlement
Southern white rhinoceros (South Africa)	Tens of thousands	10 (1900)	Some 4000	Ban on hunting, reproduction in captivity, preserves
Saiga (USSR)	Several million	Some 1500	Some 1.5 million	Ban on hunting, organization of breeding farms
Bison (Poland, USSR, GDR, FGR)	Tens of thousands	48 (1927)	More than 2000	Preserves, ban on hunting, centers for reproduction and resettlement
Reef deer (USA)	Several thousand	25–80 (1955)	600	Ban on hunting, protection of habitats
Bison (USA)	Several million	20 (1893)	More than 10,000	Protection in preserves, reacclimatization
Californian wapiti (USA)	Tens of thousands	20 (1873)	900	Ban on hunting, preserves
Abruzza chamois (Italy)	Several thousand	30 (1915)	450	Preserves, ban on hunting
White-tailed gnu (South Africa)	Tens of thousands	300 (on farm) (1900)	Several thousand	Resettlement, protection at resettlement sites
Vicugna (Peru, Chile, Bolivia, Argentina)	Some 200,000	6200 (1965)	More than 100,000	Ban on hunting and international trade, preserves, resettlement
Grey whale (Mexico, USA, USSR)	Some 20,000	Several hundred (1930's)	Some 16,000	Regulation of hunting, protection of reproduction sites

Table 11.1. *(Continued)*

Species	Population			Main restoration factors
	Initial[a]	Minimal	Current	
Koala (Australia)	Several hundred thousand	Several hundred (1937)	More than 60,000	Protection of habitats, ban on hunting
Hawaiian goose (USA)	Some 25,000	30 (1952)	900	Reproduction in captivity, reintroduction in nature
Trumpeter Swan (USA)	Tens of thousands	69 (1932)	More than 2200	Protection during migration and hibernation, establishment of reproduction centers
Great crested grebe (Britain)	Some 20,000	Some 2500 (1955)	More than 6000	Bird adaptation to nesting on artificial reservoirs
Laysan duck (USA)	Thousands	6 (1911)	More than 600	Extermination of rabbits, protection, breeding in captivity, reintroduction
Japanese crane (Japan)	Thousands	12 (1924)	More than 200	Protection of habitats, ban on hunting

[a] Prior to the overwhelming anthropogenic effect.

11.1 Protection in Natural Habitats

Two main trends in the protection of species and populations in their habitats are observed today, viz., the development of a network of protected territories and the protection of species on territories developed by man in the process of his economic activity.

11.1.1 What Areas Should Be Protected?

Insofar as it is simply impossible to protect all natural habitats, there is every reason to ask: what minimum area should be preserved to ensure the protection of the variety of species, given the low level of economic development of some territories, such as the Arctic, mountains, the greater part of the World Ocean, deserts and the northern taiga subzone in the Soviet Union. Virtually all the species can be preserved merely by way of regulating the use of economically important forms, hunting, plant collecting, etc. In the context of the considerable anthropogenic transformation of territory, the protection of specific territories is becoming increasingly important. Moreover, in case the greater part of a particular territory is changed, it is the only effective measure to preserve the variety of species.

Arguments discussed by Yablokov and Ostroumov (1983; 1985), prompt the following conclusion: nearly one third of each large physicogeographical region on our planet must be preserved in its natural state (or close to it), while living nature transformed by one third must be preserved as a background for man's economic activities. Animals and plants in the first type of territory must be ensured full protection, while those of the second category must be provided with partial, yet dependable protection. Fields, urbanized territory, roads, industry must not occupy more than 30% of dry land if we want to ward off an ecological catastrophe. If this is ensured, the preservation of the variety of species will not be unattainable.

In 1984, over 2.8% of dry land throughout the world was occupied by preserves or specially protected natural territories with similar functions. But the situation differed markedly from country to country (Table 11.2). In virtually all regions of the world the area of protected territory is rising continuously (Fig. 11.1).

The average figures cited above do not show different situations in the various parts of large countries. This can be clearly seen if the USSR is taken as an example (Table 11.3).

The conclusion that over one third of the territory of all major physicogeographical regions in the world should, on the average, be preserved in a close-to-natural state follows from the need to preserve both the variety of life (genofond protection) and the living creatures' planetary functions, such as maintaining the normal composition of the atmosphere and the quality of surface

Table 11.2. Relative area of the system of specially protected natural territories (SPNT) in some countries. (After Yablokov and Ostroumov 1983)

Country	Area: % of country's territory		Year
	SPNT System	Preserves[a]	
FRG	20.0	0.67	1981
Japan	14.0	0.20	1975
Holland	15.0	5.70	1983
Rwanda	15.0	10.43	1984
New Zealand	16.8	8.00	1981
Czechoslovakia	13.5	1.42	1982
USA	12.0	1.20	1975
Zimbabwe	11.8	6.80	1975
GDR	18.5	0.90	1984
Britain	20.2	0.56	1984
USSR	8.0	0.61	1984
Cuba	7.5	0.21	1982
Kenya	7.5	4.38	1984
France	7.5	0.70	1982
Hungary	4.63	1.01	1983
Bulgaria	1.45	0.63	1982

[a] Including those with the same environment protection regime.

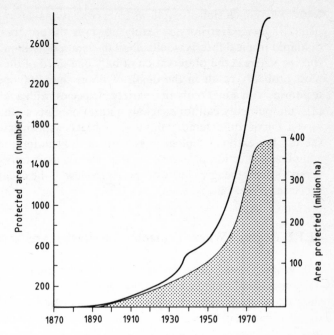

Fig. 11.1. Increase in the area and number of protected territories in the world. (Based on the 1985 United Nations List of National Parks and Protected Areas)

Table 11.3. Relative area of specially protected natural territories in the republics of the USSR

Republic	Area of	
	Preserves and natural forests	Whole SPNT system (without 1st group forests)
Estonia	2.63	6.7 (1983)
Armenia	2.61	10.6 (1984)
Turkmenistan	2.23	3.4 (1984)
Azerbaijan	1.85	3.8 (1982)
Georgia	1.73	? (1982)
Latvia	1.62	5.0 (1983)
Belorussia	1.14	4.4 (1985)[a]
Lithuania	0.73	4.7 (1983)
Uzbekistan	0.62	0.8 (1982)
Ukraine	0.88	1.2 (1982)
RSFSR	0.47	2.7 (1983)
Tajikistan	0.44	8.4 (1983)
Kirghizia	0.31	2.1 (1983)
Moldavia	0.29	1.9 (1984)
Kazakhstan	0.31	1.9 (1982)

[a] Without green zones around cities.

water. This conclusion also follows from recreational tasks. The first refers to the planet's averaged territory. In reality, however, the preservation of only one third of humid tropical forests would entail the death of millions of animal and plant species, whereas the preservation of only one third of the Arctic tundra would most probably result in the death of fewer than 100 species. It follows that territories with a markedly rich variety of species, with a markedly high "sum of life" undoubtedly call for spacious natural habitats so that this variety is preserved. On the other hand, if this is so for biotas with an extremely rich variety of species, the extreme ecological systems of the tundra and desert also call for complete protection of a considerable part of their territory to preserve their comparatively meagre choice of species in view of the peculiarities of biospheric processes there.

11.1.2 Regions Where Protected Territories Are Needed Most

The World Conservation Strategy (1980) and a number of other international documents analyzing the availability of protected territories in various regions, show that surface vertebrate species facing extinction are mostly concentrated on tropical islands, in tropical rainfall forests, marshland habitats, tropical dry and deciduous forests. Geographically, they are concentrated in the Caribbean, the West Indies, the South Pacific, Hawaii, tropical forests in Southeast Asia, on Madagascar and in Latin America.

Higher plants threatened with extinction under the pressure of anthropogenic factors are concentrated on tropical and subtropical islands (especially oceanic), in tropical rainfall forests, on arid territories, in sclerophylous forests and freshwater humid habitats (especially in Europe). In 1977, FAO's Council of Experts on Forest Genetic Resources established the following regional top priority order in organizing protected territories for such species: Africa, Southeast Asia, Australia, Mexico, the Caribbean, Central America, the eastern parts of the United States, and Canada.

The exceptionally varied ecosystems calling for urgent protection by way of establishing protected territories include primarily the tropical rainfall forests of Malaysia, Borneo, Sulawesi, Sumatra, the Philippines, New Guinea, Central and South Americas, Madagascar, mountain rainfall forests in Africa, tropical dry forests on Madagascar, sclerophylous forests in Southern Africa and West Australia, and also the island ecosystems of New Caledonia, Hawaii, the Canary and Azore islands.

A number of developing countries, where some of these large ecosystems requiring urgent protection are located, have already taken active measures to set up a network of protected territories. They include Kenya, Tanzania, Thailand, Costa Rica and Venezuela. In the past decades, the protected territories situation has swiftly been taking a turn for the better in many countries and regions. In some countries (and not a few) up to 10% of their territory remains in a close-to-natural state (e.g. Czechoslovakia).

With reference to a network of protected territories, noteworthy is the following factor related primarily to territories intensively developed by man and thus making it more difficult to establish preserves of the usual type. In some cases it seems to be more feasible to set up "spotty" preserves consisting of an archipelago of several small plots "implanted" in the developed territory. There, the protected territories may be used by animals as make-shift places for temporary stay, reproduction and rest, and also as feeding-up sites. This is what has been done on the small and the only protected plot on the Japanese island of Iriomote (the Ryukyu Islands), where the Iriomote cat (*Felis iriomotensis*) lives. By itself, the preserve cannot ensure the protection of the cat population and serves merely as a centre for carrying out nature protection measures. In all regions, broad opportunities exist for establishing such micropreserves whose area does not exceed several hectares. Especially effective are also micropreserves to protect invertebrates and small vertebrates, as well as unique nests, holes and other critical habitats.

11.1.3 What Should a Preserve Look Like?

There exists, of course, an optimal size for a preserve in each natural region. Ideally, this size should equal the territory where the population of the biggest animal and plant species can live. In this case one may say that the preserve is able to retain the entire variety of the local living population. In reality, however, it happens rarely because simple calculations show that over one million hectares of territory is required for the population of the brown bear (*Ursus arctos*). (An area of 100–200 km² is needed by a pair of brown bears to live on, while the total population should be at least several hundred species). But it is simply impossible to declare such vast territories a preserve even in the Eurasian taiga, which is a relatively less developed region. It follows that it is practically impossible to preserve the population of brown bears by establishing preserves only. Other protection measures are required, such as imposing hunting bans, setting up game reserves, and reproduction in captivity. This conclusion seems to be relevant for all large beasts of prey in the Soviet Union.

If, however, such extreme cases of ensuring the protection of large predators are excluded, it is quite possible to choose the optimal size of a protected territory. An interesting study was made in the mountain preserves of Montana USA (Table 11.4). Twenty-four game reserves ranging from 2900 to 1,160,000 ha in area were compared. In 1977–1978, 162 populations of ten big mammal species lived there: two bear species (*Ursus arctors* and *Ursus americanus*), the bison (*Bison bison*), three deer species (*Odocoileus hemionus*, *Odocoileus virginianus* and *Cervus canadensis*), the elk (*Alces alces*) and three small hoofed animal species (*Antilocapra americana*, *Ovis canadensis* and *Oreamnos americana*). Table 11.4 shows the preservation level of these species' population between 1910 and 1978.

Table 11.4. Preservation level of the populations of ten large mammal species depending on the reserve size in the mountain part of Montana (USA) from 1910 to 1978. (Picton 1979)

	Average reserve size (km²)					
	48	216	900	1834	4541	8763
Number of reserves of given size	3	5	7	3	3	3
Average number of species	4	5.4	7.7	7.7	9	8.7
Average number of biotypes	24	29.2	47.1	53.7	61	62
Percentage of preserved species	58	67	64	73	81	96
Average number of reemerged populations	0	0.2	0.7	0.3	0.5	0.7

In the plains part of the state, which has a high level of anthropogenic (mainly agricultural) land development, 45% of large mammal species have preserved. Comparison of this figure with the share of species preserved in small reserves shows that direct anthropogenic pressure on these species on the flatland was only slightly more efficient than indirect pressure through enhancing the isolation of protected territory. But the principal conclusion is this: only territories exceeding 6800 km² in area have proved to be able to sustain over 95 populations of large mammals over less than 70 years. It should be noted that, beginning in 1940, active biotechnical measures had been taken in all reserves. The data in the Table 11.4 also shows that the variety of species depends directly on the variety of their habitats.

The proponents of the theory of island biogeography took a somewhat different approach to determining the optimal size of a protected territory. Specifically, when an island's territory expands tenfold, the number of reptile species increases 50%. This and many other comparisons of animal populations on many islands have made it possible to evolve formulas based on the number of species, area and territorial variety (Fig. 11.2). It has turned out that the number

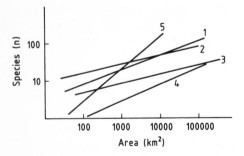

Fig. 11.2. Dependence of the number of species on the area of habitats for four groups of vertebrates on West Indies islands (*1* land birds; *2* reptiles and amphibians; *3* land mammals; *4* bats) and five land and freshwater birds on Pacific Ocean islands. (Based on data from various sources in Wilcox 1980)

of animal species (S) is inherently linked with the heterogeneity of the environment (C), which is constant for each biogeographical region, and with the island's area (A): $S = CA^z$.

The magnitude of the constant z is 0.2–0.35 for completely isolated areas and 0.11–0.17 for partially isolated ones (Diamond 1976; Simberloff and Abele 1976). In the 1970's, American zoogeographers showed that these rules are also characteristic of preserves which are a kind of island in the sea of anthropogenic landscapes. Their studies also indicated that as a preserved zone turns into an island, it becomes relatively overpopulated by species which formerly lived over a larger habitat. After some time, the number of species decreases to a level characteristic of a genuine island, or a continuous influx of species is required, as it happens on ocean islands (Diamond 1981). From this viewpoint, it is more feasible to establish preserves in groups lying relatively close to each other, so that they would be able to mutually enrich themselves. In future, when a system of reserved territories is established, this requirement is likely to be reckoned with as an important factor.

The magnitude adequate for the self-sustainment of a population is, of course, different not only for different groups but also for different environmental conditions within the natural habitat of one species. Specifically, the magnitude of various populations of sand lizard (*Lacerta agilis*) varies in different regions of the natural habitat from several hundred to 10,000 individuals. There are cases, however, when relatively small groups of hoofed animals (several dozen species) did not show any signs of degeneration over the life-span of several dozen generations. It will be recalled that, according to the Commission on Rare Species of the World Strategy for Nature Protection, the magnitude of large mammals adequate for the stable existence of a population is estimated at 500 species, of small vertebrates 5,000 species, and of invertebrates 50,000. In comparison, in 1978 in the Darwinsky Preserve (113,600 ha) situated in relatively virgin territory, there lived some 20 wolves, 60 brown bears, one pair of lynxes, 2000 wood grouse, 7000 black grouse, 2000 hazel grouse, 5000 willow grouse, and 20 eagle-owls. This data shows that the preserve's territory is unable to sustain the populations of wolves, bears, lynxes and eagle-owls.

In view of this, a territory close to one million hectares may be considered an optimal territory for preserves in countries with a temperate climate. Smaller preserves can only function if they are adjacent to large territories which are not developed intensively, or if a wise system of biotechnical measures is introduced to maintain the populations of species for which the area of protected territory proves to be inadequate.

In the forest zone and desert, the minimal territory of a preserve may be some 250,000 ha, in the moderate climate zone between 50,000 and 100,000, in the shelf zone 25,000–30,000, in the steppe zone 10,000, and in the Arctic and Subarctic between 800,000 and 1,000,000 ha (Reymers and Shtilmark 1978).

When confronted with the task of preserving the variety of species, one has to answer the question: what is better — one large territory or several smaller ones of equivalent size if combined? Extensive work, both theoretical and experimental, has been conducted to this end. Theoretical studies (Pianca 1978; Soule and Wilcox 1980 et al.) show that one large territory proves to be more reliable in ensuring protection for a great variety of species. Yet here, too, the answer is not simple, because of the "edge effect" well known in the ecology of animals and plants. This is when the basic species variety of an ecosystem is connected with the existence of transitional zones (ecotones) between different types of habitats. In the case of divided territories, however, the length of such edges may be much larger.

Choosing a territory for a preserve is often rather simple: this may be a natural habitat withe the largest variety of species or unique ecosystems, with their natural biogeocenoses disrupted least of all. In some cases, it is extremely useful to establish "critical habitats" (Ray et al. 1976). These are cases when extensively migrating species (which are many in any region, especially in the World Ocean) have to be protected (Fig. 11.3).

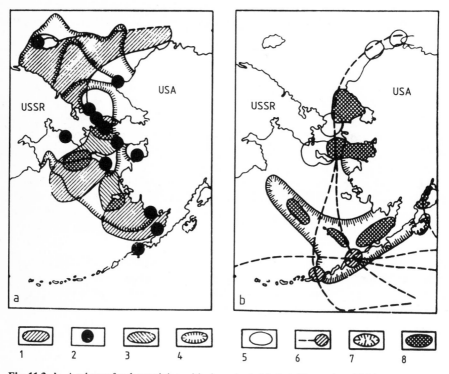

Fig. 11.3a,b. A scheme for determining critical species habitats, taking walrus (*Odobenus rosmarus*) as an example. Superposition of biologically critical hibatats (**a**) over regions subjected to especially high anthropogenic effects (**b**) makes it possible to determine more accurately those regions which require protection most. *1* regions with a permanently high concentration of species; *2* breeding-ground sites; *3* reproduction regions; *4* regions with scarce food supplies; *5* centres where the local population is actively engaged in hunting; *6* transport routes; *7* regions of intensive fishing; *8* areas where oil and gas fields are likely to be found. (After Ray et al. 1976)

The socio-economic and ecological importance of the various types of protected territories can only grow in the future. At the same time, with each passing year it will be increasingly more difficult to establish new preserves because of the growing economic development of the biosphere. The inevitable stabilization of the Earth's population coupled with a new (yet unknown in detail) food production technology will undoubtely make it possible to reduce the anthropogenic pressure on natural habitats. The present human generation is faced with the task of preserving the whole variety of life on this planet. Preserves and other protected territories have played and will continue to play a leading role in fulfilling this task, which involves the overwhelming majority of species and all ecosystems.

11.2 Captive Breeding

Captive breeding is turning rapidly into an extremely important way of preserving an increasing number of species. Breeding is captivity seems to be the only option for preserving at least several dozen animal species: they have either disappeared completely in natural habitats or the population of a species living in captivity already accounts for a considerable part of their total population (Table 11.5).

The importance of preserving species even at the cost of sustaining their population in captivity (in zoos, botanical gardens, special breeding centres, etc.) is so great that in recent years zoos and botanical gardens have increasingly been turning into centres for breeding rare species.

11.2.1 New Role of Zoos and Botanical Gardens

The idea of breeding rare animals (Lang 1972; Goodwin 1975; Martin 1975; Raczynski 1975; Wayre 1975; Hopson 1976; Durrell 1977; CAES 1979; Holden 1984) is not new. It was advanced back in 1867 by a group of scientists, the founders of the Moscow Zoo, and also in 1889 by the Smithsonian Institution when it set forth the aims of the National Zoological Park in Washington.

As of 1980, there were some 800 zoos throughout the world (30 in the USSR), where several thousand species of vertebrates and hundreds of other animal species lived.

It was in a zoo that the famous Père David's deer was saved from extinction (the animal had for several centuries been bred in one of the Emperor's parks near Beijing). After the successful reproduction of Przewalski's wild horse in a number of zoological parks in several countries, there is some confidence that this species, which has recently disappeared in nature, will not only be preserved, but also restored. In our day, zoos throughout the world have organized the breeding of

Table 11.5. Number of species of some rare mammals in captivity. (After Yablokov and Ostroumov 1985)

Species, subspecies	Number of species in captivity	Percentage of total	Year
Celebes giant civet	3	100	1981
Berbera lion	9	100	1980
Sumatra tiger	90–100	100	1982
Red wolf (USA)	46	100	1982
Père David's deer	1500	100	1985
Przewalski's wild horse	550	100	1985
Arabian oryx	316	100[a]	1979
Addax	390	100(?)	1980
European bison	2000	100 (in semi-captivity)	1982
White-tailed gnu	Some 4000	100 (including in semi-captivity)	1980
South-China tiger	Less than 50	65	1984
Amur tiger	943	78	1980
Amur leopard	42	50–60	1982
Taiwan sika deer	333	53	1980
Orangutan	603	38	1980
Lion-tamarins	330	60–70	1985
Northern white rhinoceros	13	31	1985

[a] A small amount of individuals were re-introduced to natural habitats.

many species of mammals, birds and reptiles. Moreover, the number of rare species successfully reproduced in captivity is increasing each year (Table 11.6).

The San Diego Zoological Park in California, the largest in the United States, concluded an agreement with the Government of the Malagasy Republic (Madagascar) providing for measures to organize in-captivity reproduction of certain lemur species disappearing there due to the increased economic development of the island's territory.

Table 11.6. Number of vertebrate species reproducing in captivity as of January 1, 1982. (Based on data from Olney 1984/1985, International Zoo Yearbook 1984/1985)

Group	Total Number of species	Including from Red Book of WSNP	One or both parents born in captivity
Fish	173	–	29
Amphibians	36	6	8
Reptiles	283	47	52
Birds	849	42	286
Mammals	565	120	300

Zoos should not strive to expand the variety of their animal exhibits, but rather try to ensure proper reproduction conditions for the species already living there. This is why the world's best zoos are tending to markedly decrease the variety of species and simultaneously increase the population of the residing species (Table 11.7). In 1983, as many as 75% of the animals in the Washington Zoo were born in captivity, and in 1986 more than half of the dolphins in the US marine worlds (big aquaria) were born in captivity.

All these facts show that zoos can and should be used to preserve rare species and restore their population. To this end, their material base should, of course, be markedly improved. According to the Third World Conference on Breeding Disappearing Species in Captivity (1979), the cost of preserving 2000 animal species in the world's zoos and their breeding over the coming two decades will amount to nearly US $ 25 billion. For the time being, however, the zoos largely remain the "consumers," rather than "rescuers" of rare animal species.

Botanical gardens, where tens of thousands of plant species grow, can do much for the reproduction of rare forms. There are more than 900 botanical gardens in the world, of which 220 possess systematized collections of disappearing wild flora species in various regions (Kolesnikov et al. 1977). Nearly 100 such gardens coordinate their efforts to save rare plants with the World Strategy's Special Committee on Disappearing Plants. The Arboretum of the Moscow State University's Botanical Gardens has over 800 species of trees and shrubs growing in the Northern Hemisphere, the Askania-Nova Arboretum Park 980 species, the Amur Arboretum more than 500 species of trees, shrubs and lianes, the Botanical Gardens of the Kirghiz Academy of Sciences (Frunze) 2470 forms of woody and shrubby plants.

Table 11.7. Changes in the animal population in some zoos. (After Yablokov and Ostroumov 1983)

Zoo	Number of species	Number of individuals		Year
		Total	Average per one species	
New York (Bronx)	1019	3104	3.0	1968
	598	3202	5.4	1980
San Diego	1549	4933	3.2	1968
	1050	6500	6.2	1981
Washington	452	2294	5.1	1978
	428	2481	5.8	1980
London	1548	6727	4.3	1968
	1107	8941	8.1	1980
Antwerpen	1074	5114	4.8	1968
	1161	6818	5.9	1980
Moscow	722	4161	5.8	1982
	750	5000	6.7	1985
Jersey	135	850	6.3	1978
	90	1000	11.1	1985

Some 25% or more of the local flora species are represented in the botanical gardens of Pretoria and Canberra. Specifically, in the Santa Ana Rancho Botanical Gardens in California there grow 1500 species, almost one third of the local flora, while the same number of species represent almost half the local fauna in the Botanical Gardens of the University of British Columbia (Canada). The collection of pines at the Institute of Forest Genetics in Placerville (USA) boasts 72 species, i.e. 65.4% of the species of this group growing throughout the world. According to P. Raven's estimates, by the early 1980's the world's botanical gardens had some 35,000 species, i.e. 15% of the world's flora (Raven 1981). These figures show that botanical gardens have an extremely big potential in protecting rare species. Moreover, they show that it is quite realistic to demand that all of them be represented in living collections and thus saved, at least partially, from sudden anthropogenic influences. There are cases, albeit not numerous yet instilling hope, of some disappearing forms of plants having been restored with the help of botanical gardens. In 1967, for instance, there remained as few as 23 species of the South African endemic *Orothamnus zeyheri* from the Proteacea family. By 1978, it had increased to 1139 species. The famous "living fossil" *Ginkgo biloba* has lived to our day only in Chinese and Japanese arboreti and parks, from where it has spread to many other countries. There are other higher plant species preserved in the biosphere predominantly (or exclusively) in a cultivated state. They are: *Franklinia alatamaha* (the camellia-like shrub is cultivated on the both sides of the Atlantic); *Punica protopunica* (as few as four old trees have been found growing wild ; it is cultivated in England and has spread to other regions); *Clianthus punices* (few species have survived in nature in New Zealand, yet it is cultivated extensively in gardens).

Some plant species, which have disappeared in nature completely, grow exclusively in botanical gardens. They include *Myoosotis ruscionensis* (Boraginacea); *Zea* (Euchlaena) *perennis* (Grammacea); *Lotus berthlotti* (Leguminosae); *Calandrinia feltonii* (Portulacea); *Sophora toromiro* (Leguminosae); *Bromus interruptus* (Poaceae).

11.2.2 Special Breeding and Rehabilitation Centres

Although special centres for breeding in captivity have become a topic of the environment protection literature only in recent decades, such centres for certains groups of animals have long been in existence and operate rather efficiently. These are, primarily, numerous fish-breeding facilities operating throughout the world. They preserve and multiply the population of certain salmon, sturgeon and white fish species. Specifically, the beluga and Russian sturgeon in the Azov and Caspian Seas are fully maintained thanks to artificial reproduction alone. The Caspian white salmon (*Stenodus leucichthys*) threatened with extinction (in 1970 its population dropped to 2000 producers) was saved, and 1980, the number of producers was more than 20,000, exclusively due to artificial reproduction at fish-breeding facilities. During the course of the 1980's, over 100 million fry of sturgeon fish bred at fish-breeding facilities were set free in the Caspian Sea. These and many other examples of preserving and restoring the population of species involve only those which are of practical importance for man as a source of food and other products. True, sometimes it is hard to distinguish between species having or not having practical importance for man. For example, in the 1920's, when work began in the USSR to protect the beaver (*Castor fiber*), there were as few as several hundred species. The animal was officially treated as a "natural relic," and few people believed that it could become an object of man's economic activities. Today, however, the beaver population in the Soviet Union has reached 200,000 species, and organized hunting has been practised for several years.

In the USSR, there are several special centres for breeding rare animals. They are: the Oka State Crane Breeding Centre, the Bukhara Persian Gazelle Breeding Centre, the Oka-Terrasy European Bison Breeding Centre, etc. There exist special centres for breeding rare crocodile species in India and other countries, for primates in Indonesia and Brazil, for giant tortoises on the Galapagos Islands, for musk deer and pandas in China, etc. A great many similar centres in the United States, Canada, Britain and other countries are connected with breeding rare beasts of prey and water fowl. Here are some of the largest ones: the Wildfowl Trust (Great Britain); the McDonald Raptor Research Center, McGill University (Canada); the Jersey Wildlife Preservation Trust (Great Britain) and the World Center for Birds of Prey (USA).

"Rehabilitation centres" have become a relatively new phenomenon in international practical activities aimed at saving wildlife animals. These centres provide medical assistance for wounded or diseased animals. In 1980, there were over 20 such centres in Europe alone. In the period between 1975 and 1980, ten of them treated 4179 "patients," of which 33% remained in captivity (they were unable to live in freedom) and 36% were set free following complete recuperation. A rehabilitation centre for sea mammals has for many years been operational at the San Diego Marine World. By agreement with Californian authorities, it admits all sea mammals found wounded or incapacitated on the South California Coast, such as sea lions, elephant seals, sea otters and dolphins. Many animals recuperate fully after having been injured. One killer whale, for instance, was saved and lived in captivity for 7 years, having become one of the world's best-trained whales. In Belgium, a centre for saving oil-slushed birds (Vogel Werkgroep Oostende) has been established and is now working successfully. As many as 45% of the birds coming to the centre are set free fully recuperated.

Rehabilitation centres may prove to be an important link in the system of measures to protect animals if only for the fact that they may help to sharply reduce the number of animals specially trapped to replenish zoos or for research purposes.

11.2.3 Breeding Techniques

The breeding of animals in controlled conditions makes it possible to use the latest breeding techniques, such as artificial insemination, transportation and storage of frozen sperm, transplantation of embryos, stimulation or retarding of ovulation, obtaining of many egg layings instead of one and use of other species individuals as foster parents preserve all new-born rare species individuals.

Quite new forms of animal reproduction and upkeep have emerged of late and are now developing rapidly. Formerly, rare species individuals were transported from one zoo to another for mating, whereas now artificial insemination is being used increasingly. In 1979, the first big cat (puma) was born in the London Zoo as a result of artificial insemination. In 1980, a female Amur tiger in the San Diego Zoo gave birth to a cub from a male tiger living in Hamburg. Also successful was the artificial insemination of Speke's gazelle in the Saint Louis Zoo. In 1985, an eland antelope (*Tragelaphus oryx*) in the Cincinnati Zoo gave birth to a bongo antelope (*Tragelaphus eurycerus*) as a result of a successful embryo transplant.

The foster-parent method is rather promising for breeding rare species in captivity. This technique involves, for instance, putting the eggs of one species into the nest of another species, which is not so rare, but the next of kin [e.g. the eggs of a whooping crane (*Grus americana*) are put into the nest of a sandhill crane (*Grus canadensis*), etc.]. This method is used with regard to the following species: the California condor (*Gymnogyps californianus*), the foster species being the Andean condor (*Vultur gryphus*); the masked Virginia partridge, the foster species being the Texas partridge; the black-footed ferret (*Mustela nigripes*), the foster parent being the steppe polecat (*Mustela eversmanni*). With regard to a number of large birds of prey, this technique may substantially increase the number of fly-away birds because the second nestlings of many vulturous birds usually do not survive in their own nests, due to cannibalism.

Of great importance for the organization of programmes for breeding rare animals in controlled conditions is a worldwide account of rare species individuals, their registration in pedigree books and the working out, on this basis, of plans of interbreeding schemes to minimize the consequences of inbreeding (see Chaps. 4, 3). This real management of genetic resources is already in effect practically on a global scale with regard to Przewalski's wild horse, Amur tiger, oryx, Indian rhinoceros, golden marmoset, okapi, and 53 other species.

11.3 Domestication and Cultivation

In the 16th-17th centuries, there were merely several dozen species of house plants, whereas now their number exceeds several thousand. Tens of fish species and other vertebrates have been domesticated. When man cultivates plants or domesticates animals, he controls and supports their life.

The breeding (not simply keeping) rare animals in captivity or cultivating rare plant species for economic or aesthetic purposes makes one feel confident that they become part of man's household and are thus saved from full extinction as species. However sharp their decrease in nature may be in future, they will not face complete extinction and their genetic pool will be at least partly preserved.

Yet this involves its own difficulties. One is the problem of preserving the genetic diversity of initial species because in the process of domestication all species take on new features and their genetic pool changes. It seems that in the context of the alternative: "to preserve a species in a domesticated form or to lose it forever," there can only be one desirable solution, viz., to preserve it even in a changed form (as is the case, for instance, with the European tur, whose genetic pool has been partly preserved in our present cattle).

Let us consider, albeit briefly, the prospects opening up in this sphere of living nature protection.

The first stage of domestication is farm breeding for animals and planting for plants (for some species this may be the last stage of domestication in the foreseeable future). The scale of animal farm breeding is already extensive. In the period between 1969 and 1979, animal farms in the RSFSR produced more than 100 million foxes, sables, polar foxes, American minks, chinchillas, swamp beavers, etc. Annually, over 13 million fur-bearing animals are bred in cages in the USSR.

Big successes have been scored in breeding a number of mammal species and turning them into semi-domesticated animals. One of the most graphic examples is the Soviet method of breeding sables (*Martes zibellina*) in captivity, which was developed back in the 1930's. The Soviet Union has achieved good results in domesticating eland antelopes (*Tragelaphus oryx*), European elks (*Alces alces*), Japanese deer (*Cervus nippon*), and European red deer (*Cervus elaphus*). The domestication of the eland antelope was first begun in Russia in the Askania-Nova Preserve. Soviet animal farms also successfully breed nilgais (*Boselaphus tragocamelus*), brindled gnus (*Connochaetus taurinus*), zebras (*Equus zebra*, *Equus burchelli*, *Equus grevyi*), and bisons (*Bison bison*, *Bison bonasus*).

In some regions, the breeding of wildlife hoofed animals has proved to be more profitable than the raising of traditional cattle. This becomes more understandable if one takes into account the fact that buffalos, for example, graze on grass 26 cm high, zebras some 20 cm, antelopes gnus some 12 cm, and rhinoceros as low as 7 cm. It is clear, therefore, that one and the same section of the African savanna may serve as pasture for a varied community of wildlife hoofed animals, whose aggregate biomass exceeds that of household cattle. It is not surprising, therefore, that 80% of land in the central part of Zimbabwe is used today for the farm breeding of wildwife animals. In 1984, in South Africa there were nearly 10,000 commercial animal farms breeding eland antelopes (*Tragelaphus oryx*), brindled gnus (*Connochaetus gnou*), bonteboks (*Damaliscus dorcas*), springbucks (*Antidorcas marsupialis*), impalas (*Aepyceros melampus*), greater kudus (*Tragelaphus strepsiceros*), oryxes (*Oryx gazella*), 300,000 sika antelopes, and also tens of thousands of introduced European red deer (*Cervus elaphus*) and fallow deer (*Cervus dama*).

In the early 1970's, some 500,000 springbucks lived on such farms. In 1985, in Australia there were nearly 200 farms breeding 12,000 fallow deer (*Cervus dama*) for commercial purposes.

In the mid-1970's, in the United States there were several hundred farms breeding exotic game animals, such as European fallow deer (*Cervus dama*), mouflons (*Ovis ammon musimon*), axis deer (*Cervus axis*), sika deer (*Cervus nippon*), blackbucks (*Antilope cervicapra*), nilgais (*Boselaphus*), barbary sheep (*Ammotragus lervia*), and also some 25 other big mammal species, including giraffes, lamas and zebras. In the state of Texas alone, the total number of farm-bred rare mammals was increasing each year:

1963	1975	1979
13 species, 13,000 animals	34 species, some 50,000 animals	51 species, more than 72,000 animals

This trend is very promising because many of these species possess a set of truly remarkable properties. Specifically, the springbuck is well adapted to the desert conditions of South Africa, feeding on succulents and plants poisonous for domestic cattle and being able (like the oryx) to do without water for several days. The female springbuck puts on 80% of its adult weight over the first 7 months of her life. The meat of wildlife hoofed animals has high gustatory qualities, and eland milk has long been used for medicinal purposes.

The capybara (*Hydrochoerus capybara*) in Venezuela and the Gambian rat (*Cricetomys gambianus*) in Nigeria are two interesting examples of relatively small mammals domesticated to obtain food. In the 1980's, farms breeding Eurasian hedgehogs (*Erinaceus europaeus*) have appeared in Britain. Their "produce" enjoys increasing demand among gardeners: domesticated hedgehogs protect gardens from snails.

Some bird species (ostriches, wild turkeys, partridges, quails and certain other wild birds) are bred on farms for commercial purposes.

Reptile farming is also gaining momentum. When the natural resources of venomous snakes had been exhausted, several snake centers in the USSR began breeding adders (*Vipera berus*) for commercial purposes.

The problem of reptile farm breeding (tortoises, snakes, crocodiles, lizards) is not as simple as it may seem. It often happens that the damage done to natural populations by taking some animals (or egg layings) for farm breeding is higher than the benefit received from reproducing young animals on farms. Such reproduction in captivity cannot, of course, be a reasonable alternative to hunting. This is why, beginning in 1980, animals "born in captivity" are only those whose parents have mated in artificial conditions. (In this case even rare species trading is allowed).

The domestication process has so far involved only a small number of invertebrate species. One can refer to the domestication of several species of honeybee (*Apis mellifera*), silkworm (*Bombyx mori*), Indian lac insects (*Ericerus pela*) in Southeast Asia and cochineal insects (*Porphyrophora hameli*), which, prior to the spread of chemical dyes, gave one of the best natural pigments. In recent decades, the list has greatly expanded due to the industrial reproduction of several insect species used in the biological struggle (predatory ladybirds, ichneumon flies, etc.).

In recent decades, some unusual objects of planting and farm breeding have appeared, such as earth worms. Thousands of tons of such worms are produced in the Philippines and some states in the USA to be used not only as a fishing bait, but also as feed for the industrial breeding of shrimps, eels and frogs. Rubber-tree, cacao-tree and palm-tree plantations have proved to be good places for the farm breeding of the giant African snail kalutara (*Achatina fulica*) that gives some 200 g of delicious meat. The farm breeding of dozens of species of tropical butterflies for collectors and souvenir-makers is widespread in Brazil, Taiwan, Papua New Guinea, while in Britain they breed aboriginal butterflies for their preservation (Underwood 1983).

New animal and plant species are constantly being broadly involved in man's economic activities within the framework of aquaculture (see Chap. 10.2). Account should also be made of the breeding of many fish, singing birds and other species by amateurs for aesthetic purposes. The population of some fish species living in home aquaria may well exceed their population in nature by many times.

With regard to planting, two aspects of the problem should be emphasized from the viewpoint of the protection of living nature. First, if made a crop, the plant has good chances of survival and the preservation of its genofond (similarly to the domestication of animals which are becoming rare because of uncontrolled hunting or other economic activities). Several thousand medicinal plants (some 0.5% of the world fauna species) are already used in Pharmacopeia. Almost all plants possess specific secondary metabolites with a pharmacological potential. More than 30% of drugs in the world are made of some 500 species of medicinal plants, which are partially cultivated. In the USSR alone, over 50 species of drug plants are grown on 26 specialized farms. In Hungary, 160 such species are cultivated on 60 farms. Dozens of plant species are used in perfumery and other industries as technical raw materials. Thousands of plant species, including several species that have disappeared in nature (e.g. *Ginkgo biloba*, certain palm trees and sequoia), are cultivated as decorative plants. Many forms of food and oil plants are also known as cultivated species only.

Here is but one of the recent examples. Several countries have begun to cultivate jojoba (*Simmondia chinensis*), a Mexican plant known some 10 years ago only to a narrow circle of researchers and local people. It has been found that jojoba seeds produce valuable technical oil, practically analogous to cachalot spermaceti (the cachalot population in the World Ocean has sharply decreased as a result of excessive whaling).

To sum up, the future of several thousand plant species may not cause special worries if only for the fact that they are needed by man for his economic activities and are thus "within the field of his vision". In general, however, although many animal species have already been introduced into the economy and domesticated (or are on the way to domestication in the broad sense of the word), there are still broad prospects in this field. The process of domestication will undoubtedly help preserve the variety of living nature.

11.4 Ecological Engineering

Purposeful practical measures taken to solve problems in the field of the protection of living nature through the use of ecological methods and approaches may be called ecological engineering (by analogy with genetical engineering in molecular biology).

The development and implementation of scientifically grounded acclimatization and re-acclimatization projects are the simplest forms of ecological engineering.

11.4.1 Acclimatization for Saving

In order to ensure more dependable protection of not numerous representatives of rare species, they can be resettled in localities less susceptible to anthropogenic influences, which they could not reach during the course of natural settlement. Remote or relatively isolated islands are often chosen as such natural reserves of living nature.

Islands may be regarded as specific natural evolutionary laboratories which can be used by man for saving certain forms of life. Resettlement of certain disappearing species onto islands with an environment suitable for their life could, in our opinion, be conducive to the creation of some sort of natural continental islands (if, of course, one feels confident that the new settlers will not cause death to any endemics). Thus, a group of axis deer (*Axis calamiensis*) – local animals that have become dangerously few — were resettled on the island of Keyleyvit (Philippines). The European mink (*Mustela lutreola*), driven out by the American mink (*Mustela vison*), which has been spreading rapidly throughout Eurasia over recent decades, was resettled on the island of Kunashir (Kuril Islands).

Quite successful was the resettlement of the famous Madagascar lemur aye-aye (*Daubentonia madagascariensis*), the only representative of the aye-aye family. By 1969, as few as some 50 aye-ayes were left in nature. In 1963, on the initiative of enthusiasts, several aye-ayes were transported to the island of Nozi

Mangaba near the northern coast of Madagascar. As a result, in 1983 the first baby aye-aye was seen on the island.

This pertains to any isolated habitats suitable for the life of species becoming rare in their natural habitats. Water reservoirs and isolated mountain ranges may also serve such "islands". Specifically, the disappearing gavial (*Gavialis gangeticas*) and marsh crocodile (*Crocodylus palustris*) were introduced into isolated lakes in Pakistan, where they had never lived because of their remoteness.

11.4.2 Reintroduction

The reintroduction of animals and plants which have once disappeared, into their original habitats is being practised ever more extensively. The last Arabian oryx (*Oryx leucoryx*) was killed in Central Oman in 1972. In 1979–1983, 41 oryxes out of the "world oryx herd" composed on the initiative of the World Wildlife Fund and the World Strategy for Nature Protection by way of registering all the oryxes living in captivity throughout the world, were reintroduced into Oman and Jordan. In 1981, the first new-born oryxes were seen in the Oman herd. The brown argus butterfly (*Lycaena dispar batavus*) disappeared in Britain's fauna as far back as 1851 due to large-scale marsh reclamation and burning-down of vegetation. The attempts to reintroduce this subspecies failed four times: they were all ecologically unsound. Success came as late as 1927. During the first years following the reintroduction, the caterpillars were bred in fish-wells, and *Rumex hydrolapatum*, one of the sorrel species and the butterfly's main feeding plant, was planted in localities where it was set free. As a result, the subspecies was completely restored in Britain a few years later.

The Atlantic salmon (*Salmo salar*), which had disappeared in the Rhine in 1958, was reintroduced in 1983. The lynx (*Felis lynx*), which disappeared in the fauna of many European countries in the 18th-19th centuries, has by now been successfully reintroduced in Sweden, West Germany, Austria, Yugoslavia, and France (1983). True, it would be far-fetched to speak of fully-fledged, self-sustaining populations of this species in Western Europe, yet lynx groups will most probably continue to exist there safely for decades to come at the deme level. Following that, special biotechnical (in fact, ecological engineering) measures will be required to make them viable. The white stork (Ciconia ciconia) disappeared in Switzerland in 1950. As a result of the successful reintroduction of the bird from West Germany and France, 79 stork pairs made nests in Switzerland in 1979, having hatched 131 nestlings all in all.

Table 11.8 lists several reintroduction projects, both accomplished and under way.

In all the cases of successful reintroduction, extensive ecological work was first done, involving both the creation of suitable habitats and the preparation of animals for resettlement. For example, the reintroduction of lynxes, which were

Table 11.8. Some examples of successful reintroduction of disappearing species

Species	Where from	Where to	Year
Blackbird (*Acridotheres tristis*)	Britain (Jersey)	Bali Island	1984
Mauritius rose pigeon (*Nesoenas mayeri*)	Britain, Jersey	Mauritius	1985
Golden lion tamarin (*Leontopithecus rosalia*)	USA	Brazil	1983
Père David's deer (*Elaphurus davidianus*)	Britain	China	1985 1960
Kit fox (*Vulpes velox*)	USA	Canada	1984
Arabian oryx (*Oryx leucoryx*)	Several countries	Oman, Jordan	1979– 1983
Brown argus butterfly (*Licaena dispar*)	Holland	Britain	1927
Hawaiian goose (*Branta sandvicensis*)	Britain	Hawaii	1962
Two pheasant species	Britain	Taiwan	1960's
Raven (*Corvus corax*)	FRG	Holland	1969– 1972
Fallow deer (*Cervus dama*)	FRG	Cyprus	1980– 1982
Alpine ibex (*Capra ibex*)	FRG	Alps	1970's
Alpine chamois (*Rupicapra rupicapra*)	FRG	Alps	1970's

partially taken from Czechoslovakian zoos, called for a set of measures to teach the animals how to live in the natural environment.

11.4.3 Changing Migration Routes and Hibernation Sites

Swedish scientists have recently carried through a very interesting ecological-engineering project. The population of lesser white-fronted geese (*Anser erythropus*) in Sweden had been drastically reduced from 30,000 in 1953 to 200–300 in 1983 as a result of mass shooting at their hibernation sites in Turkey. Several years ago, eggs of some lesser white-fronted geese were put in the nests of barnacle geese (*Branta leucopsis*), which hibernate in relative safety in Holland and are plentiful in Sweden. As a result, in 1983 5 adult and 20 young lesser white-fronted geese, which had come back after their hibernation in Holland, were seen for the first time in Stockholm (Fig. 11.4). The whooping crane (*Grus americana*), whose population is extremely sparse, nests in the Canadian Subarctic and hibernates on the Texas coast (USA). Two successful attempts to set up new populations of this species were undertaken. One egg was taken out of each nest in the Wood Buffalo National Park and put in the nests of more populous species. As a result, several

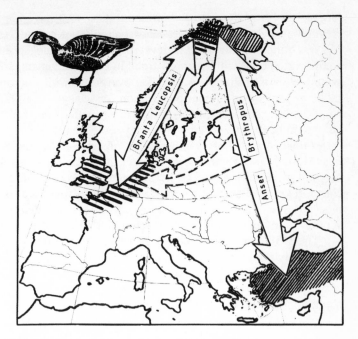

Fig. 11.4. A diagram showing changes in the hibernation sites of the lesser white-fronted goose (*Anser erythropus*) of the Swedish population. *1* the nesting and hibernation habitats of the lesser white-fronted goose; *2* the nesting and hibernation habitats of the barnacle goose (*Branta leucopsis*); *3* migration routes. (Yablokov and Ostroumov 1985)

dozen birds grew up safely amongst their foster parents and began to hibernate together in New Mexico, spending the summer in Idaho and Wyoming.

A similar technique involving a change in migration routes is now being employed with regard to some rare cranes in the context of a joint Soviet-American research programme.

11.4.4 Development of New Biogenocenoses

As man develops and reconstructs the biosphere, he sometimes sets up new ecosystems to replace those completely destroyed, which, in longer perspective, will take on all the basic features of natural ecosystems (cyclical self-regeneration, high homeostasis, etc.).

Most of Europe, for instance, is covered with secondary forests. Heath waste grounds and most of the other shrubbery habitats (chaparral, etc.), which are viewed today as natural ecosystems, have all the features of semi-natural eco-systems. The creation of permanent water reservoirs on the basis of the irrigation system in Central Asia has enriched the variety of local fauna. Not only migrant bird nesting sites but also stable hibernation sites have appeared there. The creation of such artificial biogeocenoses should not be underestimated.

It is quite possible that the development of new biogeocenoses with pre-set complex functions will become a trend in ecological engineering. Thus, special swamped plots or shallow-water lakes overgrown with macrophytes are created today to provide additional purification for sewage. Phosphorus and nitrogen compounds in sewage water stimulate the growth of water plants. This artificial ecosystem of the lake-swamp type may serve as an excellent habitat for many water fowl species. For example, the Lyubertsy and Lyublino irrigation fields near Moscow are habitats for sandpipers, gulls, partridges, ducks, ruffs and owls. Migrant birds, such as storks, herons, geese and cranes make stopovers there. Hares, foxes, hedgehogs, polecats, ermines and weasels also live there. These biocenoses have formed spontaneously, without man's deliberate participation, yet they point to a possible way of action in the field of ecological engineering.

The USSR and other industrialized countries often recultivate territories temporarily used as building sites, occupied by mines or turned into industrial waste lands. In some regions such anthropogenic waste lands account for up to 2% of a country's territory. In future, it might be useful to develop water-plant plantations (microphytes) which would perform three functions simultaneously: (1) recycling some wastes from urbanized territories; (2) serving as habitats for water fowl; and (3) shouldering part of the recreational load. All these are real reserves for creating islands of life in the biosphere, especially in regions which need them most.

11.5 Genetic Banks

If all the reserves of preserving a species in situ have been exhausted, one must at least think of the possible preservation of its genofond in the form of seeds, sex cells, etc. in special depositories — banks. There are more than 40 seed banks throughout the world today. They are part of the depository system of the International Bureau on Plant Genetic Resources established by the 1972 Stockholm Conference on the Environment. The banks contain over one million species of agricultural plants.

The seed bank does not, however, solve the problem of preserving the genofond of all plants because many of them reproduce vegetatively only. By now, scientists have developed methods for conserving the plant genome by deep freezing tissues at growth points, embryonic structures, sex and somatic cells. The preservation of meristems seems to be most important for genome preservation: it is the meristems which make it possible fully to restore and reproduce the given genotype. In principle, this method of preserving plant genofond (Fig. 11.5) is probably applicable to all species, although so far it has been developed in detail with regard to only a few of them.

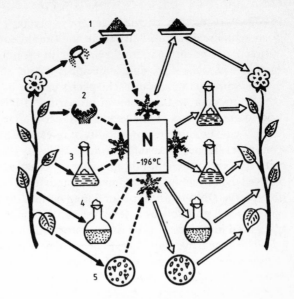

Fig. 11.5. A general diagram of a bank of plant cells and tissues. *1* pollen collection; *2* meristem isolation; *3* obtaining a culture in a nutrient medium; *4* a culture of cells in suspension; *5* a culture of embryoides; *6* the *dotted arrows* show programmed freezing after adding cryoprotectors; *center* storage in liquid nitrogen; *right* recultivation after defreezing and washing the cryoprotectors and subsequent full regeneration of plants. (Butenko and Popova 1979)

Back in the 1960's, the first banks of microorganisms were established. But the purpose was not to preserve genofond per se, but to conduct experiments and ensure the safe storage of the pathogens of particularly hazardous infections. With regard to prokaryotes, the establishment of a bank seems to be a practical task even today.

The establishment of animal genetic banks is much more complicated. In the 1960's, the first banks of producer bulls' and roosters' sperm appeared. Considerable differences in the animal sex cells' sensitivity to freezing, storing and defreezing hold out little hope that simple and universal techniques will be developed to store the genes of disappearing animal species. Bull sperm, for example, can be stored in frozen form for decades, whereas horse and sheep sperm can only be kept for a few hours. Besides, it has turned out that non-fertilized animal egg cells suffer especially from freezing. At the same time it has been established that the embryos of such mammals as cows, sheep, rabbits and mice can be stored, if deep-frozen in the early stage of development, for a long time.

Scientists have developed a master scheme for preserving and restoring animals from conserved sex and somatic cells, zygotes, gonads and embryos (Veprintsev and Rott 1979), where all separate stages have been performed on various subjects. The programme includes inter-specific nucleus transplants, androgenesis and gynogenesis, the inter-specific transplantation of zygotes and gonads, and does not contain any major scientific limitation to its realization. It is safe to say that the scheme in general is practical, although it is clearly still far from actual realization. This conclusion should not, however, delay the conservation of disappearing species' genomes. In the not-so-distant future, molecular biology will make it possible to decipher the genetic code of any species. Then, genetic information about each species can be recorded in the form of tables containing data on the chemical properties of DNA molecules. The complete nucleotide sequence of the genomes of certain viruses has already been established, which offers a real possibility of re-establishing these forms in laboratory conditions. Hence some infinitesimal possibility that the fixation of living tissues with fixers (e.g. alcohol-propionic acid), which do not destroy the irreversibly native DNA structure and suppress the activity of nucleases, will make it possible to preserve DNA of any type for a long period. If this possibility is realized, our progeny will probably be able to re-establish, on the basis of the DNA structure preserved in a piece of fixed tissue, forms which are bound to perish.

The gene bank may prove to be extremely important even today to keep up a numerically small group of animals facing degeneration as a result of inbreeding. The danger of the latter's negative influence is real for rare species bred in captivity. It is known, for instance, that the fertility of Przewalski's horse dropped sharply after several generations had been bred in captivity.

Apart from material difficulties, there are no major obstacles to prevent a system of genetic banks for plants, animals, fungi and microorganisms from repeatedly duplicating in future the entire wealth of the world's genetic resources and from serving as a reserve, the "last straw" for restoring lost species if the worst comes to the worst. But to make this possibility a reality, considerable material expenses and strenuous research are needed.

11.6 Regulated Evolution

Evolutionarily, the extinction of species, irreversible changes in populations under the impact of anthropogenic factors and other numerous changes in the genetic structure of species and biogeocenoses are, in fact, normal processes. Species either change, adapting themselves to new anthropogenic conditions, or die out. If the anthropogenic transformation of the biosphere proceeded more slowly in a way commensurate with the usual duration of the evolutionary process (it takes, on the average, nearly one million years for a new species to emerge and live for several years as one generation), most of the existing species could probably adapt themselves to new anthropogenic conditions and develop relevant adaptive properties. This is what happened in Africa where the emerging genus *Homo* had coexisted with the Pleistocene fauna for a long time (at least three million years). That time proved to be sufficient to develop certain adaptive properties by virtually all the representatives of that fauna (elephants, hoofed animals, birds of prey, etc.), which has ensured the survival of the African

Pleistocene fauna up to the present. In all the other regions of the world the Pleistocene fauna was destroyed during the course of just a few millennia following the appearance of man (see Chap. 4.1).

Studying the laws that govern the development of living nature, Man discovered the principle of artificial selection and grew, through selection, a great many forms hitherto non-existent in nature. The number of varieties of basic cultivated plants, such as wheat, rice, corn, bananas and cotton, runs into hundreds or even thousands in separate cases (there are, for example, over 4000 varieties of wheat). No less numerous are the species of ornamental plants, such as tulips, gladioli and roses. The variety of domesticated animal breeds is fewer in absolute figures, yet it is also rather high. Specifically, there are some 1500 strains of cattle and several hundred breeds of pigs, sheep, chickens, dogs and cats. It is no exaggeration to say that Man has created several dozen thousand species hitherto non-existent in nature by regulating the evolutionary process.

This form-making explosion has occurred in the setting of a sharp decrease in the variety of wildlife forms. Unfortunately, the process of anthropogenic form-making cannot compensate for the shrinkage of natural variety, so dangerous to the biosphere. All the new forms hitherto developed by man were intended exclusively to satisfy his utilitarian or aesthetic needs.

Today humanity must think seriously about ways to regulate evolutionary processes from the biospheric, ecosystemic positions. Hitherto this regulation has been narrowly anthropocentric, although in the final analysis the biospheric, ecosystemic regulation is also anthropocentric because it involves the survival and well-being of mankind on Earth.

On the basis of this, we may say that regulated evolution is man's conscious and purposeful development of living nature in order to enhance the efficiency of the utilization of natural resources in the near future and maintain the stability of the basic characteristics of the biosphere and the ecosystems of large regions. Similarly to natural biological evolution, this process should be accompanied by changes in the genotype composition of populations, the development of adaptive abilities, the emergence and extinction of species (although not on today's mass scale) and the transformation of ecosystems.

Let us consider, albeit briefly, two major trends in regulated evolution: (1) changing (adjusting) wildlife species for living in the biosphere changed by man; and (2) regulating the development of biogeocenoses.

11.6.1 Intentional Change of Wildlife Species

We have already referred to the coexistence of relatively large animal species (badgers, beavers, martens, etc.) and man on urbanized territories (see Chap. 10.5.). In all such cases, the species have gone through the process of natural selection resulting in the survival of those not afraid of Man. With regard to beavers, this selection was effected on the Popelno Farm of the Institute of Genetics and Animal Breeding of the Polish Academy of Sciences.

All such examples serve as models for processes which in the foreseeable future might involve species living on urbanized territories. In a more distant perspective, as the biosphere will change ever increasingly, such processes might also spread to the populations of the other species or most of them. It is difficult to speak of the concrete forms of this process today. Possibly, it will differ somewhat with regard to animals living on urbanized territories ("in man's shadow," as it were) and those preserved on natural territories.

As to the first category of animals, man may take special measures to maintain their populations. The examples are many even today: squirrels, badgers, martens, weasels, polecats, ducks, geese, starlings, storks, eiders, certain gull species in Western and Eastern Europe; skunks and some foxes in the United States; some monkeys, mongooses, flying foxes and some herons in Southeast Asia.

As to the second category, various forms of habitat amelioration can be applied. Indicative is the analysis of how artificial nesting sites may be used in forests. It has been shown that in some cases wood ducks (*Aix sponsa*), screech owls (*Otus asio*), golden-winged woodpecker (*Colaptes auratus*), Carolina sparrows (*Troglodites*), common opossums (*Didelphis marsupialis*), western grey squirrels (*Sciurus griseus*), eastern fox squirrels (*Sciurus niger*) and northern flying squirrels (*Glaucomys sabrinus*) are more willing to use artificial nesting sites than natural hollows (McComb and Noble 1981).

12   Scientific Foundations and Theory
of Living Nature Protection
Within the System of Biological Science

The study of various problems of the protection of living nature by various disciplines, such as genetics, ecology and biogeography has resulted in the emergence of the theory of living nature protection, which makes a scientific basis to a newly applied or science-production trend, viz., the biology of the protection of nature or conservation biology (Soulé and Wilcox 1980).

12.1 Biological Sciences and the Scientific Foundations
of Living Nature Protection

This book has shown the many nature conservation problems confronting systematics and biogeography, zoology and botanics, development biology and biochemistry. Here we will try to show the relevance of more active studies in this field from the viewpoint of separate disciplines, taking some special problems of living nature protection as examples.

12.1.1 Systematics. Zoology and Botanics

The development of the systematics of plants, animals, fungi and prokaryotes is vitally important for the organization of living nature protection on this planet, a fact that is increasingly understood today by many people, not only biologists. Unfortunately, we are far from being in a position to describe everything living on our planet: according to various estimates, the 1.5 million species of living organisms known today account for merely from 50 to 10% of the total variety of nature.

It will be no exaggeration to say that systematics is one of the most backward fields in biological sciences related to the protection of living nature. The description of the variety of forms in nature should have preceded the rapid development of ecology, molecular biology and other branches of modern biology. This has not happened, however, and serious efforts are needed today to rectify the current absurd situation.

The species problem should primarily be singled out of the special problems of systematics connected with the protection of living nature. Although this problem has been developed rather fully by the theory of evolution, the application of the chief species criterion, viz., the inability to interbreed in natural conditions (a species as a genetically stable system), is impossible in practical systematics either now, or in the future. There are also difficulties in determining objectively the taxonomic ranks of a group (subspecies or species, genus or family, etc.). The importance of these theoretical problems for the practical protection of nature is evident inasmuch as the taxonomic rank of a group is a most important criterion for determining the priority of action.

12.1.2 Ecology, Biogeocenology

It is still not clear what regularities govern the development and interaction of biogeocenoses — the "bricks" that form all large-scale ecosystems and the biosphere as a whole.

The problems of biogeocenology connected with certain perspective trends in the protection of living nature include the development of a theory of regulation of the development of artificial biogeocenoses (semi-natural and anthropogenic complexes) in agriculture (agrobiocenoses) and other economic spheres (urbanized and recultivated territories, ruderal biogeocenoses, etc.).

It seems reasonable to mention several ecology's underdeveloped fields. Here are only three of them. First, development of the theory and methods of ecological prognostication seems to be the most serious and pressing task of the special branches of ecology dealing with the protection of living nature. Of special importance are the applied aspects of such prognostication connected, on the one hand, with the management of utilized living natural resources and, on the other, with the study of the future of species and other forms sharply decreasing numerically and facing extinction.

Second, the study of the stability of biological systems is a general task of ecology, important for the theory of nature conservation. It is still unclear what mechanisms ensure the stability characteristic for living systems (e.g. Regier and Cowell 1972; Stugren 1986).

Third, the study of populations and species as elements of ecosystems is a field of ecological research which is important for the protection of nature and which is unjustifiably given little attention. This includes the study of obligatory and specific links between species. It is necessary to have a clear idea of how many and which species are obligatorily dependent on the given rare or disappearing species. This alone will make it possible to determine the damage the biosphere will suffer from its possible extinction. Groups of species within which at least one species is absolutely necessary for the stable existence of another (or other) species could be called ecological clusters (see e.g. Ostroumov 1986). We believe that it is relevant to study the biology of disappearing species and ecosystems from this viewpoint as well.

12.1.3 Biochemistry

The role of biochemistry, molecular biology and chemistry and their contribution to the solution of nature conservation problems is more important than is commonly believed (see Chapt. 2).

At least four major problems can be singled out here. The first involves the study of chemical interactions between living organisms. These interactions seem to play an immense role in regulating many processes that occur in populations and biogeocenoses, but our knowledge of them is, insufficient (for a survey see Harborne 1982; Ostroumov 1984, 1986; Schlee 1986). These interactions may be especially vulnerable to anthropogenic influences, such as pollution. The unsuccessful attempts to forecast the behaviour of ecosystems and to control them can stem from the failure to duly take them into account. Knowledge of the role of concrete pheromones, attractants, phago repellents and colines (the participants in allelopathic interactions between plants), as well as that of various other ecological chemoregulators and chemomediators (Ostroumov 1984; 1986) in general will help reduce the use of poisonous chemicals in agriculture and forestry.

The second problem is the study of the scale and regularities of the bioaccumulation of pollutants in living organisms (e.g. Johansen 1981; Drossos et al. 1982; Drasch 1983; Oliver and Niimi 1983; Geyez et al. 1985). This task is extremely hard to fulfil methodically because of the variety of taxons. Yet it is necessary for the following reasons. The presence of pollutants, including toxic, in organisms' tissues not only leads to their own pathology but also destroys the populations of higher trophic levels, including Man and domestic animals. On the other hand, the rapid concentration of certain pollutants in individual animal or plant species makes it possible to use the latter as "biosensors" of ecosystems' pollution. Finally, the large-scale accumulation of pollutants in certain organisms (e.g. plants) allows us to think about their possible use to clean polluted water and soils. Such plants could be grown there and then removed, with the biomass having "sucked out," as it were, the water and soil pollutants (see e.g. Weinberger and Greenhalgh 1985).

The third problem is the study of mechanisms through which pollutants affect living organisms. This problem has so far been studied quite inadequately (e.g. Richardson and Waid 1982; Leo 1984; Sentzova and Maximov 1985 and other works). The study of mutagenic effects was given much more attention than that of the membranothropic and enzomothropic (proteinothropic) effects of chemical pollutants. Further studies in this field will show whether it will be possible to forecast, on the basis of the chemical structure of a substance, its biochemical effect and the level of its ecological danger to the biosphere (see also Wexler 1988; Kolb Meyers 1988).

The fourth problem is the study of the biotransformation of pollutants (e.g. Alexander 1981; Atlas 1981; Wood 1982; Miyamoto et al. 1988). Results obtained so far indicate that pollutants in an organism may transform in two directions. First, they may develop a lower level of toxicity and persistence. In this case it might be interesting to study the regularities of the degradation of a particular pollutant in order to speed up the process. The most degradable classes of chemical substances could be singled out and used to replace more toxic and stable components discharged into the biosphere. Second, pollutants may transform into more toxic (e.g. more mutagenous or carcinogenic) and more stable compounds. In this case it is important to find ways to block such processes. It is desirable that non-degradable or mutagenous substances should be recognized more surely on the basis of their chemical composition. Toxic substances in an organism or living cells may be rendered harmless by binding them and excluding from metabolism (e.g. proteins-metallothyoneins).

These problems related simultaneously to ecology, biochemistry, chemistry and the protection of the biosphere could be dealt with by a branch of science called biochemical ecology (Ostroumov 1986) or ecological biochemistry (Harborne 1982; Schlee 1986).

12.1.4 Genetics

The study of laws governing the change and formation of the genetic diversity is also important for developing a strategy for the protection of living nature.

The major genetic problems directly related to the protection of living nature include determining the real danger of the mutagenous effect of various anthropogenic factors on living organisms and preserving the necessary level of changeability when the population of a species decreases and its structure is disrupted. These problems also include lowering the inbreeding effect when breeding rare species in captivity or when the population of a species in nature is numerically limited (the so-called problem of minimal numbers), as well as the study of the normal level of genetic heterogeneity, i.e. the level characteristic of various populations, species and species groups.

The study of the first of these problems will make it possible to have a better idea of the relative danger of various pollutants, this knowledge being a basis for improving the habitats. The solution of the second group of problems will allow us to better organize the system of rare species interbreeding in breeding centres and zoos throughout the world, to choose optimal strategies when making territories populated by not numerous large animals into protected zones, to determine more accurately the dangerously low level of the population of a particular species.

12.1.5 Developmental Biology

There is no need to prove especially the importance of studying the embryotoxic and teratogenic effects of various pollutants. This field of research also includes studies (not very active so far) dealing with the peculiarities of the local impact of various substances on the growth and development of individual organs and systems of various groups of organisms.

In principle, researchers have developed programmes for preserving sperm, embryos, individual cells and their cultures, etc. for a lengthy period (decades and centuries) for the gene bank. Implementation of such programmes is an important field of research in the sphere of developmental biology.

12.1.6 Other Sciences

The physiology of animals and plants is important for the protection of living nature, just as the entire range of physiological sciences. Of great value is the study of the effects made on the organism, its systems and tissues by various anthropogenic factors, both physical and chemical. Physiological sciences are also important for nature conservation in connection with the intensification of agriculture. Certain problems of plant physiology should be resolved in order to introduce phytodrome plant growing and markedly improve agrochemistry with regard to the use of fertilizers in traditional agriculture.

Radical improvements in the hormonal regulation of reproduction and population are certainly important for the physiology of animals and man, for the optimization of man-nature relations and for an active demographic policy.

Microbiology may make an important contribution to the transformation of pollutants in biocenoses, to the formation of optimal cenoses in purification facilities (e.g. Mudrack and Kunst 1988) and to the ecologization of a number of industries (the use of ferments and immobilized microorganisms, the development of biotechnology, etc.) and energetics.

Palaeontology can use its specific methods of research to reconstruct past biotas. This in turn can reveal today's trends in faunistic and floristic change, especially as regards data obtained over the recent geological periods of our planet's development. This may prove important both for forecasting the overall ecological situation and for determining the level of anthropogenic changes in various regions.

The emergence and disappearance of species and other forms in nature is the subject matter of the evolution theory. The evolutionary approach is absolutely necessary to resolve most of the practical tasks of living nature protection — from the prevention of certain forms from extinction by way of developing specific adaptive properties enabling them to live side by side with man in a changed biosphere to the regulation of the development (i.e. evolution) of biogeocenoses. The concept of "regulated evolution" advanced back in the 1920's by Vavilov remains a symbol of faith today (Vavilov 1937).

The importance and role of certain elementary evolutionary factors, such as the mutation process, magnitude waves, isolation and natural selection, with regard to living nature protection still await their thorough analysis. Any such factor can be determining at a particular stage of a species' existence in a biosphere changed by man. For example, the rate of the mutation process may prove inadequate in a swiftly changing environment. The range of population variations may become a limiting factor when preserving populations that are not numerous. Isolation may not take on its "classical" property of a factor augmenting evolutionary differences, but rather become a factor accelerating the extinction of disintegrated species during the course of natural habitat insularization (disintegration is caused by the break of inter-population links).

12.2 Axiomatization of the Theoretical Foundations of Living Nature Protection

Generalization of single facts leads to the formation of ideas about the subject matter of a study, while further accumulation of facts makes it possible to develop a concept and put forward certain hypotheses. In a narrow scientific sense, a theory is not merely theorizing about this or that subject matter. It is rather a single hypothesis picked out of a number of others and proved not to be contradictory to all known facts. A theory greatly speeds up both further penetration into the essence of a phenomenon and the formulation of practical applications. This is why the development of a theory is justly considered as a criterion for the development level of a particular field of knowledge.

So far, there is no theory of the protection of living nature. But there are many groups of facts, systematized to a varying degree, and a number of working hypotheses. In view of the intensive work in this direction in all biological sciences and many related disciplines, it is quite possible that the development of such a theory is a matter of a not-so-distant future.

Summing up some intermediate results is extremely useful at any stage of theoretical work. In the briefest possible form, such results could be presented as a system of conjectural axioms. A variant of the axiomatization of knowledge in the field of living nature protection is given.

From the viewpoint of practical activities in the field of nature protection, the propositions in the right column are axioms which are to be taken into account when taking relevant measures.

To make the system of axioms fuller and complete, the following two important axioms must be added:

1. The protection of living nature is the necessary element of the stable and harmonious socio-economic development of both individual countries and human society at large. Nature protection goals must be taken into account at any stage of economic development and nature utilization.

Scientific Propositions	Conclusions
1. Life exists in the form of communities of living organisms in a certain space in the form of biogeocenoses. Biogeocenoses are the elementary units of the biosphere.	The preservation of biogeocenoses and other ecosystems is necessary to maintain life and the functioning of the biosphere.
2. The qualitative variety of living nature is the foundation for maintaining the composition of the most important biospheric systems: soil, the atmosphere, fresh and sea water.	The preservation of the qualitative variety of living nature is necessary to maintain the habitat parameters vital for man.
3. The main specific features of living organisms are: the capacity for self-preservation through self-reproduction with changes (convariant reduplication) and the capacity for self-regulation.	For life to be preserved on the Earth, anthropogenic factors must not hinder the realization of these chief properties of life.
4. Each biological species is a unique result of evolution; each species has a unique genetic pool.	All genetic pool in the biosphere has a potential value and is to be protected. The preservation of the qualitative variety of living nature is in keeping with mankind's supreme interests. It ensures the continuous expansion of the spectrum of benefits received from nature, including the potentially unlimited expansion of the various uses of natural products in various industries, medicine and culture.
5. Species in nature live closely tied in with other species, i.e. in the form of ecological clusters (associations of connected species).	The preservation of one species without the preservation of other species connected with it (ecological clusters) is impossible in nature. To preserve individual species, it is necessary to preserve the ecosystems the species in question belong to.
6. Each species is a system of evolutionarily integrated populations. A species cannot exist for long if represented by a single isolated population.	For an individual species to be preserved over a lengthy period, it is necessary to preserve (or set up) an integrated system of populations.

Scientific Propositions	Conclusions
7. Due to inevitable variations in their magnitude, populations with a small number of species will sooner or later be affected by the fluctuation leading to their extinction.	Populations with a small.number of species are always under the threat of extinction and require special protection measures.
8. Food production involves multiple accumulation of many chemical substances (bioaccumulation), with many such substances undergoing substantial changes in living organisms and ecosystems (biotransformation).	The bioaccumulation of pollutants makes even the insignificant levels of pollution potentially dangerous to living nature and man. The biotransformation of substances may lead to the toxification (increased toxicity) of initially harmless compounds.

2. Anthropogenic effects on the biosphere involve actions by specific people. This is why to preserve the biosphere and its components, measures should be taken to ensure that the awareness of the great value of the biosphere and all living species and of the importance of their preservation should be high on the list of values and moral principles of any person, especially those who make decisions.

The above variant of the system of axioms does not seek to embrace the entire variety of life's properties important for the organization of nature protection. This variant may need to be tied in with a system of broader generalizations, primarily the "rules" of ecology established by Commoner (1971), the "iron laws of nature protection" set forth in 1980 by Ehrlich and possibly other theoretical propositions of a fundamental nature.

The rules formulated by Commoner (1971): (1) "Everything Is Connected to Everything Else," (2) "Everything Must Go Somewhere," (3) "Nature Knows Best," and (4) "There is no such thing as a free lunch."

The laws given by Ehrlich (1980), with some changes:

1. In nature protection, there can only be successful defenses or retreats. An offensive or advance is impossible because destroyed species and ecosystems cannot be restored.
2. Population growth and nature protection contradict each other, and they are incompatible.
3. The economic system's growth mania and nature protection are incompatible in principle.
4. Making decisions on the utilization of the Earth in order to reach only short-term, immediate benefit of *Homo sapiens* is deadly dangerous not

only to all living nonhuman organisms, but in the long run also to mankind.
5. Nature protection is a matter of the well-being and survival of mankind.

These 5 "iron laws" are slightly modified after Ehrlich (1980).

12.3 Conclusion

The protection of living nature plays a special role in resolving a cardinal problem of our time, viz., the preservation of the conditions for the development of mankind. This book has analyzed some major aspects of this problem and specified and systematized some principal solutions.

The development of the biosphere was a continuous change of the specific composition of living organisms — biogeocenoses and large ecosystems. But the change has never been so swift as it is today (the dying-out of pangolins in the late Cretaceous period lasted for several million years). Moreover, formerly the variety of living species (the sum of life) increased rather than decreased as a result of major changes, this being the main feature of life evolution on our planet. On the one hand, an increase in the number of developing species increasingly varied adaptability to an increasingly changing environment and stimulated further evolution. On the other, this meant an increase in the variety and complexity of biogeocenoses and other larger ecosystems and the biosphere as a whole.

As a result of active photosynthesis, oxygen accumulated in the Earth's atmosphere, forming the ozone screen protecting the planet's surface from the Sun's intensive UV radiation. The life processes of ocean organisms facilitated the stabilization of the chemical composition of water and the formation of soils. This made the biosphere more stable and protected it from the effects of outer space.

The situation began to change with the advent of Man and the rapid development of human society. Up to this day, mankind has been solving its problems without caring much about the future of the biosphere. In the 20th century, the limitation and exhaustibility of the entire range of the Earth's living resources has become evident not only to a narrow circle of scientists but also to laymen.

Neither the air composition, nor the quality of freshwater, nor the fertility of land can be maintained and ensure life without the proper operation of natural biogeocenoses. This is why the protection of living nature is also the protection of the conditions for man's vital activities. Living nature with all its variety is an irreplaceable component of our life. By his very nature, Man will always remain a biological being tied in with the multiform world of living nature through billions of links of which he is not always aware. But the protracted change of generations often does not allow Man to comprehend in time the danger of particular anthropogenic effects, whose variety and potency are increasing at a breath-taking speed.

The invaluable world of living nature is now in a critical state resulting from man's activities (the analysis of relevant problems is given in Part I of this book). Living nature is falling into a decline at a menacing speed. The rate of species loss throughout the world seems to be several species per day and continues to increase. As a result, if the current trend persists, up to 20% of the genetic pool may be lost within the next few decades. It will include a considerable number of species yet unknown to science (among them those potentially important for medicine and biotechnology) and many forms which could become sources of new agricultural plants and domestic animals. This will be an irreplaceable loss of real and yet unknown benefits.

The rate of the anthropogenic destruction of species reflects the immense scale of the destruction of natural ecosystems. This destruction has already resulted in a sharp decrease in the stability and productivity of agriculture and forestry (including hunting and fishing), a decline in the sanitary and epidemiologic state and general destabilization in some regions of the biosphere. We may well assume that in the future the functioning of the biosphere at large will be disrupted, despite the fact that the considerable inertia of biospheric processes conceals many irregularities and stresses that gradually accumulate in the biosphere.

The loss of the genetic pool, the destruction of large ecosystems and the disruption of the biosphere irreversibly limits the choices of ways of society's further development (Fig. 12.1). The preservation of living nature will increasingly become a limiting factor of society's development.

The anthropogenic decline of the biosphere, the pecular degeneration of evolution into its opposite — the involution (curtailment) of life — contradicts the

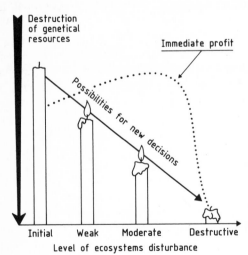

Fig. 12.1. A diagram of correlation between the level of genetic diversity preservation, the degradation of natural ecosystems, the immediate benefits received by man, and the possibility of choosing development directions. (Yablokov and Ostroumov 1983)

chief natural-historical regularity in the development of the biosphere, viz., a continuous increase in the sum of life on this planet. Man's power over nature is limited to his ability to cognize its laws and use them in his activities. The social form of matter's motion (human society) cannot abrogate biological laws, just as the biological form of matter's motion does not invalidate the laws of physics and chemistry. By ignoring these laws, mankind destroys the very foundation of its existence.

Is there any hope that the variety of life put at the disposal of man will be preserved on our planet? In our opinion, the answer is "yes" — so far. However, there are several prerequisites for the preservation of this variety.

The first is connected with the fundamental conclusion that the human race can coexist with any form of life. Being a product of the biosphere, man cannot be antagonistic to living nature. It is only important to establish a correct relationship between the biosphere's resources and society's activities, to organize the "economy of life." This economy should be based on the unique ability of living organisms to multiply in a geometrical progression, combined with their inherent changeability. In nature, these properties serve as the mainspring of the process of natural selection and evolution as a whole. In the future biosphere, which will be controlled by man, these properties must be realized not only to further increase the sum of life, but also to obtain food and other necessary resources.

The second prerequisite for preserving the qualitative variety of life is the experience rapidly accumulated by man in preserving and managing natural resources. Substantial results have been achieved in the varied relevant activities described in Part II of this book.

The following quite realistic perspectives emerge: transition from hunting and fishing to economic management with regard to an increasing number of species; expansion of the spectrum of species and forms of living organisms used in economy; reduction and elimination of the adverse effects suffered by living nature from modern agricultural technology, industry, energetics and transport; expansion of natural protected territories; improvement of nature protection legislation and ecologization of consciousness; establishment of centres for saving and breeding species which have become almost fully extinct and their restoration in natural habitats; and measures to make the utilization of living nature inexhaustible. In a longer perspective, this will also include discoveries of new benefits concealed in living nature and development, on their basis, of a biotechnology which will be able to revolutionize entire industries. In a still longer perspective, this will also involve an ever broader switch-over to natural raw materials reproduced by living nature and full ecologization of technologies and virtually all spheres of society's life.

The preservation of living nature calls for joint efforts to be made by every person in particular and mankind in general. Everyone must "think globally and act locally." The community of people and each of its members is held responsible. The awareness of this responsibility elevates ecological propositions within the system of our philosophical values still higher.

So, what is needed is the ecologization both of Man's consciousness and society's functions, including the economy at large, industry, agriculture, etc. This ecologization has already begun. It means Man's advancement to a new position is respect to living nature which corresponds better to his long-term interests. It means elaboration of a strategy of Man's coexistence with living nature, the coevolution of man and the biosphere.

Against the background of all mankind's vital problems, including ecological problems, the absurdity of militarism becomes especially evident. The arms race gives rise to specialized industries consuming vast human, financial and natural resources and polluting and destroying ecosystems and species habitats. The arms race stimulates urbanization and exacerbates the ensuing problems. Tense inter-state relations invariably narrow the range of possibilities for active cooperation in the protection of both individual species and the large regions which need that protection most. This is happening at a time when it has become clear that unprecedented international cooperation is needed to preserve a healthy environment and living nature, its pivotal component.

It has been established that if a frog is thrown into hot water it reacts instantly and may leap out of the vessel. If, however, the same frog is put in cold water which is then heated gradually, it will die unwittingly. The fear that mankind may find itself in the same situation has prompted us to write this book. More than any other specialists, biologists who study the biosphere see many end results of the unprecedented anthropogenic change of the environment, to which we all are party. In the entire known Universe, there is nothing even remotely resembling the biosphere of our planet. The survival of mankind depends entirely on the state and operation of the biosphere. The data discussed in this book clearly indicate that the biosphere is vulnerable to anthropogenic influences at all levels. Our activities are such that the natural capacity of the biosphere for reparation is nearing its limit, while in many cases the limit has been reached or even passed.

Care for environmental protection (primarily, for the preservation of the qualitative variety of living nature) is not an obstacle to, but rather the necessary condition and vital prerequisite for society's further progress.

The ecologization of all aspects of society's activity is a key feature of the biosphere's transition into the noosphere[1]. Succeeding generations will consider it a crime if we continue our activities, ignoring their long-term ecological consequences and above all their effects on the Earth's living nature. If, having read this book, the reader agrees with these conclusions, we will consider our first mission accomplished. If, moreover, the reader starts thinking about his or her own activities in the light of what has been said, our second and main mission will also be accomplished.

The ecologization of human thought as a planetary phenomenon is probably one of the most common features of the natural, social and technical sciences of the last third of the 20th century. If this trend continues and expands, this may be

---

[1] From the Greek noos (νoos) — intellect, reason, sense, thought.

the solution of some problems of the protection of living nature. In any ecosystem, there are upper levels which concentrate the flows of matter and energy. Similarly, in the protection of the environment, the preservation of life is the key problem of the protection of nature in general. The preservation of the variety of species and the biogeocenotic qualitative variety of life on our planet is the basis of the protection of life. However complex the relevant tasks may be, this problem must be solved as a condition for mankind's further progress.

Bibliography

Ackefors H (1977) Production of fish and other animals in the sea. Ambio 6:192

Adjarov D, Ivanov E, Keremidchiev D (1982) Gamma-glutamyl transferase: a sensitive marker in experimental hexachlorobenzene intoxication. Toxicology 23:73–77

Agarwal A (1979) Pesticide resistance on the increase, says UNEP. Nature 279:280

Ahsanullah M, Edwards RRC, Kay DG, Negilski D (1982) Acute toxicity to the crab *Paragrapsus quadridentatus* (H. Milne Edwards) of Kuwait light crude oil, BR/AB dispersant, and oil-dispersant mixture. Aust J Mar Freshwater Res 33:459–464

Ajayi SS (1976) Observations on the biology, domestication and reproductive performance of the African giant rat Cricetomys gambianus Waternouse in Nigeria. Mammalia 39:343–364

Akbar S, Rogers L (1985) Effects of DDT on photosynthetic electron flow in Secale species. Phytochemistry 24:2785–2789

Alexander M (1981) Biodegradation of chemicals of environmental concern. Science 211:132–138

Alexandrov Y, Golubovsky MD (1983) A role for viruses and exogenous DNA in natural mutagenesis. Genetica (The Hague) 19:1818–1827 (in Russian)

Allen R (1978) Living marine resources. Sourcebook for a World Conservation strategy. Second Draft. IUCN, Morges, pp 1–27

Allen R, Prescott-Allen C (1978) Threatened vertebrates. Sourcebook for a World Conservation Strategy. IUCN, Morges, pp 1–17

Allendorf F (1988) Conservation biology of fisches. Conservation Biology 2:145–148

Amante LC (1970) The National League against the destruction of birds, Italy. Biol Conserv 2:1–2

Anderson A (1985) War on war on migrating birds. Nature 318:6

Anderson RC (1977) The value of roadside verges to wildlife. Vic Res 19:14–15

Andrews M (1977) The lives that live on man. Tapliner, New York, pp 1–183

Anon (1986) Acid rain envoys urge coal-cleaning technologies. Chem Eng News 64:5

Anonym (1976) Endangered species: when to act. Sci News 110:266

Anonym (1983) This project does not need West Africa's wild chimps. AWJ Quart 32:6

Asferg T, Jeppesen JL, Aaris-Sørensen J (1977) Graevlingen (Meles meles) og graevlingejagten i Dannmark 1972/73. Dan Vildtunders Vildtbiol Stat Kalø 28:1–56

Ashmore M, Bell N, Rutter J (1985) The role of ozone in forest damage in West Germany. Ambio 14:81–87

Ashton FM, Crafts AS (1981) Mode of action of herbicides. Wiley, New York, 525 pp

Atlas RM (1981) Microbial degradation of petroleum hydrocarbons: an environmental perspective. Microbiol Rev 45:180–209

Auckley J (1982) Can the red wolf be saved? Conservationist 43:15–16

Aureli P, Dracos A, Lorch L Von, Zanasi F (1975) Impiego di mutanti antibiocoresistenti per l'identificazione della natura dell'agente antimicrobico repertato in campioni di latte crudo. Mondo Latte 29:648–654

Austin B (1985) Antibiotic pollution from fish farms: effects on aquatic microflora. Microbiol Sci 2:113–117

Avise JC, Crawford RL (1981) A matter of lights and death. Nat Hist 90:6–15

AWJ (1980) No call for poisons where dogs guard sheep. Animal Welfare. Inst Inform Rep 29:3

Azoulay-Dupuis E, Bouley G, Blayo MC (1982) Effects of sulfur dioxide on resistance to bacterial infection in mice. Environ Res 29:312–319

Babich H, Stotzky G (1985) Heavy metal toxicity to microbe-mediated ecological processes. Environ Res 36:111–137

Bach W (1985) Forest dieback. Experientia (Basel) 41:1095–1104

Baher JM (1976) Marine ecology and oil pollution. Barking Appl Sci

Baidya KN (1986) The firewood crisis in India: a major socio-cultural problem of rural communities. Int J Environ Stud 26:279–294

Baker J, Schofield C (1985) Acidification impacts on fish populations: a review. In: Adams D, Page W (eds) Acid deposition. Plenum, New York, pp 183–222

Baker JM (1981) The investigation of oil industry influences on tropical marine ecosystems. Mar Pullut Bull 12:6–10

Baker JM (1983) Impact of oil pollution on living resources. Environmentalist 3:1–48

Baker M (1980) Conservation + cooperation = 2 nol. Chance for the tule elk. Zoonooz 53:6–9

Balk F, Kieman JH (1984) Future hazards from pesticide use. IUCN Comiss on Ecology, Paper N 6, 100 pp

Bandiera S, Sawyer T, Campbell M, Fujita T, Safe S (1983) Competitive binding to the cytosolic 2,3,7,8-tetrachlorodibenzo-p-dioxin receptor. Biochem Pharmacol 32:3808–3813

Barnet D (1980) Premier rencontre nationale des centres de réhabilitation pour animaux sauvages. Homme Oiseau 18:239–241

Baroyan OV (1982) Nature of human health. In: Zemlja ljudei (Men's Earth, in Russian). Znanie, Moscow, pp 35–49

Baroyan OV, Lepikhov AM (1975) Why conserve it? Man Nat 8:7–83 (in Russian)

Barrett RT (1979) Small oil spill kills 10–20,000 seabirds in North Norway. Mar Pollut Bull 10:253–255

Barthova J, Kalasova H, Pacakova V, Leblova S (1985) The effect of s-triazine-type pesticides and chlorinated hydrocarbons on lactate dehydrogenase. Environ Res 36:26–31

Barus V, Lusk S, Gajdusek J (1981) Fauna ryb a jeji zachovani v Československu. Památky Přir 6:619–623

Batten LA (1972) Breeding bird species diversity in relation to increasing urbanisation. Bird Study 19:157–166

Beary JF (1979) Mercury in sperm whale meat. Science 206:1260

Bedrohte Pflanzen und Tiere (1980) Naturschutz Naturparke N 98:43–44

Begley S, Hager M, Carey J (1984) A cancer "epidemic in fish. Newsweek 20 Feb, p 66

Benirschke K (1977) Genetic management. Int Zoo Yearb 17:50–60

Berger AJ (1979) Hawaii's birds. Nature Conserv News 29:9–13

Berglund S, Davis RD, L'Hermite P (eds) (1984) Utilisation of sewage sludge on land: rates of application and long-term effects of metals. Reidel Dordrecht, p 231

Bernard A, Lauwerys R (1984) Cadmium in human population. Experientia (Basel) 40:143–152

Berry RJ (1976) Inheritance and Natural history. Collins, L, p 350

Biali F (1983) L'élevage du gibier. Bull Techn Inf Min Agr, N 377–378, num spec: "Chasse et agr": 243–257

Bialobok S (1973) Sources of danger and the protection of plants in the immediate surroundings of man. In: Protection of Man's Natural environment, Wars. PWN, pp 180–197

Biessman A (1982) Effects of PCBs on gonads, sex hormone balance and reproduction processes of japanese quail Coturnix coturnix after ingestion during sexual maturation. Environ Pollut (Series A) 27:15–30

Bilder RB (1980) International law and natural resources policies. Nat Res J 20:451–486

Bird DM, Lague PC, Buckland RB (1976) Artificial insemination vs. natural mating in captive American kestrels. Can J Zool 54:1183–1191

Bird D, Peakall D, Miller D (1983) Enzymatic changes in the oviduct associated with DDE-induced eggshell thinning in the kestrel. *Falco sparverius*. Bull Environ Contam Toxicol 31:22–24

Biswas MR, Biswas AK (1978) Loss of production soil. Int J Environ Stud 12:189–197

Blab J, Nowak E, Trautmann W, Sukopp H (1984) Rote Liste der gefärdeten Tiere und Pflanzen in der Bundesrepublik Deutschland. Kilda, Greven, p 270

Black JJ, Holmes M, Zapisek WF, Dymerski PP (1980) Fish tumor pathology and aromatic hydrocarbon pollution in a Great Lakes estuary. In: Hydrocarbons and halogenated hydrocarbons in an aquatic environment. Proc Int Symp Anal Hydrocarbons and Halogenated Aquatic Environ. Ontario, 1978. New York, London, pp 559-565

Black JJ, Evans ED, Harshbarger JC, Zeigel RF (1982) Epizotic neoplasms in fishes from a lake polluted by copper mining wastes. J Nat Cancer Inst 69:915

Blagosklonov KN (1980) Aviafauna of the big city and possibilities of its transformation. In: Ecology, geography and conservation of birds. Nauka press, Leningrad, pp 144-155 (in Russian)

Bleavins MR, Aulerich RJ, Ringer RK, Bell TG (1982) Excessive nail growth in the European ferret induced by aroclor 1242. Arch Environ Contam Toxicol 11:305-312

Bloesch M (1980) Drei Jahrzehnte schweizerischer Storchansiedlungsversuch (*Ciconia ciconia*) in Altreu, 1948-1979. Ornithol Beob 77:167-194

Bloom H, Smythe L (1984) Environmental lead and its control in Australia. Search (Syd) 14:315-318

Boch MS, Mazing VV (1979) Ecosystems of peatlands of the USSR. Nauka press, Leningrad, p 187 (in Russian)

Bogovski S, Kangur M, Kadakas V (1982) Tumors in fish and their connection with pollution of water. In: Problems of modern ecology: ecological aspects of environmental protection in Estonia. Tartu, p 114 (in Russian)

Bothe M (1980) Transfrontier environmental management. IUCN Environ Policy Law Pap N 15:391-401

Boudou A, Desmazes J, Georgescauld D (1982) Fluorescence quenching study of mercury compounds and liposome interactions. Ecotoxicol Environ Saf 6:379-387

Bourdeau P, Treshow M (1978) Ecosystem response to pollution. In: Butler GC (ed) Principles of Ecotoxicology SCOPE 12. Wiley, New York

Boyden S (1980) Ecological study of human settlements. Nature ressour 16:2-9

Brädy RF, Tobias T, Eagles PF, Ohrner R, Micak J, Veale B, Dorney RS (1979) A typology for the urban ecosystem and its relationship to larger biogeographical landscape units. Urban Ecol 4:11-28

Breitender K (1979) Ein europäisches Netzwerk des Naturschutzes wurde geschaffen. Umweltschutz 17:351-354

Breitender K (1980) Ein europäisches Netzwerk des Naturschutzes wurde geschaffen. Umweltschutz 18:15-17

Briedermann L, Stubbe C (1979) Im Darwin-Naturschutzgebiet. Unsere Jagd 29:302-303

Brodie PF, Pasche AJ (1982) Density-dependent condition and energetics of marine mammal populations in multispecies fisheries management. In: Mercer MC (ed) Multispecies approaches to fisheries management advice. Can Spec Publ Fish Aquat Sci 59:35-38

Brosselin M (1980) Le massacre des migrateurs. Homme Oiseau 18:163-166

Brown L (1976) Saving birds of prey. Wildlife (Lond) 18:414-419

Brown LR, Durning A, Flavin C et al. (1989) State of the World 1989. WW Norton, New York, London, 256 pp

Browning JA (1974) Relevance of knowledge about natural ecosystems to development of pest management program for agro-ecosystems. Proc Am Phytopathol Soc, pp.191-199

Buff K, Brundl A, Berndt J (1982) Differential effects of environmental chemicals on liposome bilayers. Fluorescence polarization and pesticide-lipid association studies. Biochim Biophys Acta 688:93-100

Bunyard P (1986a) Waldsterben and the death of Europe's trees. Ecologist 16:2-4

Bunyard P (1986b) The death of the trees. Ecologist 16:4-14

Burdin KS (1985) The fundamentals of biological monitoring. Moscow University Press, Moscow, p 158

Burger J (1981) Behavioural responses of herring gulls *Larus argentatus* to aircraft noise. Environ Pollut A24:177-184

Buringh P (1980) Limits to the productive capacity of the biosphere. In: Future sources organic Raw Material. CHEMRAWN I. Invit Lect World Conf Toronto, 1978. Oxford ea, pp 325-332

Buringh P (1985) The land resource for agriculture. Philos Trans R Soc Lond B Biol Sci 310:145-325

Burkhard WD (1982) Bleivergiftete Wasservögel am Untersee. Natur Mensch (Nuernb) 24:282-286

Burton R (1983) Coping with marine pollution. Fairplay Int Shipp Weekly 287:15-16

Butenko RG, Popova AS (1979) Bank of plant cells — prospects and problems. Priroda 4:2-11 (in Russian)

Butler PA (1973) Residues in fish, wildlife and estuaries. Pest Monit J 6:238

Byczkowsky J, Sorenson J (1984) Effects of metal compounds on mitochondrial functions: a review. Sci Total Environ 37:133–162

Cabridenc R (1985) Degradation by microorganisms in soil and water. In: Sheehan P, Korte F, Klein W, Bourdeau P (eds) Appraisal of tests to predict the environmental behavior of chemicals. Wiley, New York, pp 213–232

CAES (1979) The center for acting on endangered species. Annu Rep, pp 1–4

Caldwell J (1985) Novel xenobiotic conjugates. Biochem Soc Trans 13:852–854

Caldwell J, Paulson G (eds) (1984) Foreign compound metabolism. Taylor & Francis, London, Philadelphia, p 328

Caufield C (1980) The threat to Britain's wild places. New Sci 88:204–205

Cerovsky J (1985) Plant conservation in the GDR. Threatened Plants Newsletter 15:20–21

Cerovsky J, Holub J, Prochazka F (1979) Cerveny seznam flory ČSR. Pamatky Prir 6:361–363

Chanda JJ, Anderson HA, Glamb RW, Lomatsch DL, Wolff MS, Woorhees JJ, Selikohf IJ (1982) Cutaneous effects of exposure to polybrominated biphenyls (PBBs): the Michigan PBB incident. Environ Res 29:97–108

Chappie M, Lave L (1982) The health effects of air pollution: a reanalysis. J Urban Econ 12:346–376

Cherfas L (1980) Still no ban on green turtle imports. New Sci 85:988

Chetverikov SS (1983) Waves of the life. In: Chetverikov SS Problems of general biology and genetics. Nauka, Novosibirsk, pp 76–83

Chown M (1986) Smoking out the facts of nuclear winter. New Sci 1538:24–25

Clark DR, Krynitsky AJ (1983) DDE in brown and white fat of hibernating bats. Environ Pollut A 31:287–299

Clarson TV, Nordberg GF, Sager PR (1985) Reproductive and developmental toxicity of metals. Scand J Work Environ Health 11:145–154

Clausing G, Grun P, Beitz H (1980) Möglichkeiten zur Untersuchung und Vermeidung der Beeinträchtigung der Vogelwelt durch Pflanzenschutzmittel. Nachrichten Pflanzenschutz DDR 34:139–143

Clutton-Brock J (1976) The historical background to the domestication of animals. Int Zoo Yearb 16:240–244

Cole HA (1979) Marine nature reserves. Mar Pollut Bull 10:364

Cole HA (1980) Changes in North Atlantic fisheries. Mar Pollut Bull 11:31

Collins M (1985) Tarantula sales under scrutiny. Traffic Bull 7:5–8

Commoner B (1971) The closing circle. Alfred A. Knof. NY p 326

Connell D, Miller G (1984) Chemistry and ecotoxicology of pollution. Wiley, New york, p 444

Conti P, Gigante G, Alesse E, Cifone M, Fieschi C, Reale M, Angeletti P (1985) A role for Ca^{2+} in the effects of very low frequency electromagnetic field on blastogenesis of human lymphocytes. FEBS Lett 181:28–32

Convention on International Trade in Endangered Species of Wild Fauna and Flora (CITES) signed in 1973. In: Nilsson G (ed) The endangered species handbook. Animal Welfare Institute. Washington, 1983, pp 204–209

Conway W (1980) An overview of captive propagation. In: Soule ME, Wilcox BA (eds) Conservation Biology. Sinauer. Sunderland, pp 199–208

Coppinger R, Smith C (1983) The domestication of evolution. Environ Conserv 10:283–292

Corbett R, Wright K, Bailie A (1984) The biochemical mode of action of pesticides. Academic Press, London, p 382

Cornwallis RK (1969) Farming and wildlife conservation in England and Wales. Biol Conserv 1: No 2

Corwin T (1980) The economics of pollution control in Japan. Environ Sci Technol 14:154–157

Cory L, Fjeld P, Serat W (1971) Environmental DDT and genetics of natural populations. Nature 229:128–130

Cramer HH (1967) Plant protection and world food production. Pflanzenschutz-Nachr 20:1

Cramp S (1977) The problem facing birds of prey. The view of ornithologists and conservationists. World Conf Bird of Pray, Vienna, Oct 1975. Pert Proc Basingstoke, pp 9–11

Crawford M (1985) Condor agreement reached. Science 229:1248

Cronk QCB (1980) Extinction and survival of endemic vascular flora of Ascension Island. Biol Conserv 17:207–219

Crubinsky Z, Gawloska J, Zabieravski K (1977) Rezerwaty przyrody w Polsce. Panstwowe Wydawnitctwo Naukowe, Warsaw, pp 528

Cruz AA de la (1982) Effects of oil on phytoplankton metabolism in natural and experimental estuarine ponds. Mar Environ Res 7:257–263

Culliney J (1976) The forest of the sea. Life and death of continental shelf. San Francisco, Sierra club books, p 290

Curry-Lindahl K, Harroy JP (1972) National parks of the world, vol 2. Africa, Asia, Australia and Oceania. Golden, New York, p 240

Cushing DH (1975) Fisheries resources of the sea and their management. Oxford Univ Press, p 95

Cushing DH (1980) European fisheries. Mar Pollut Bull 11:311–315

Dagani R (1980) Aquatic toxicology matures, gains importance. C & EN, June 30:18–23

Dathe H (1972) Standesämter für das Wildtier. Urania (DDR) 48:24–25

Davis SD, Droop SJM, Gregerson P et al. (1986) Plants in danger. IUCN, p 450

Davison AW, Bailey IF (1982) SO_2 pollution reduces the freezing resistance of ryegrass. Nature 297:400–402

Declaration. The world campaign for the biopshere (1982) Environ Conserv 9:91–92

Degobbis D, Smodlaka N, Pojed I, Skrivanic A, Precali R (1979) Increased eutrophication of the Northern Adriatic sea. Mar Pollut Bull 10:298

Demayo A, Taylor MC, Taylor KW, Hodson P (1982) Toxic effects of lead compounds on human health, aquatic life, wildlife plants, and livestock. CRC Crit Rev Environ Contr 12:257–305

Deutsch E (1983) Elektrische Stühle für Vögel. Vogelschutz 1:11–13

Dezhkin VV (1983) World hunting and game management. Lesnaya promyschlennost Moscow, pp 360 (in Russian)

Diamond AW (1981) Reserves as oceanic islands: lessons for conserving some East African montane forests. Afr J Ecol 19:21–25

Diamond JM (1976) Island biogeography and conservation: strategy and limitation. Science 193:1027–1029

Diamond JM, May RM (1976) Island biogeography and design of natural reserves. In: May RM (ed) Theoretical ecology. Saunders, Philadelphia, pp 163–186

Dipple A (1983) Formation, metabolism and mechanism of action of polycyclic aromatic hydrocarbons. Cancer Res 43:2422–2425

Dolan R, Goodell H (1986) Sinking sites. Am Sci 74:38–46

Donat P, Sedlacek K (1982) Kriteria pro hodnoceni ornitofauny a Cerveny seznam ohrožených druhu ptaku v ČSSR. Památky Přir 7:423–438

Doughty R, Choban A (1977) Texas exotics at home on the range. Wildlife 19:204–208

Douglas JH (1975) Harvesting of wild. Sci News 107:259

Dover M, Croft B (1985) Pests' increasing resistance to pesticides: a growing health and economic problem. Ambio 14:64–65

Drasch GA (1983) An increase of cadmium body burden for this century — an investigation on human tissues. Sci Total Environ 26:111–119

Dreistadt S, Dahlsten D (1986) Medfly eradication in California: lessons from the field. Environment 28:18–44

Drossos CG, Mavroidis KT, Papandopolou-Daifotis Z, Michalodimitrakis DN, Salamalikis LX, Gounaris AK, Varonos DD (1982) Environmental lead pollution in Greece. Am Ind Hyg Assoc 43:796–798

Duinker JC, Hillebrand MTJ, Notting RF (1979) Organochlorines and metals in harbour seals (Dutch Wadden Sea). ICES, CM 1979/E: 40, Mar Pollut Bull 10:360

Dunford C, Poore D (1978) Drylands. IUCN, p 33

Dunnett GM (1979) Nature conservation in the marine environment. Mar Pollut Bull 10:318

Dupuy PM (1980) International liability for transfrontier pollution. IUCN Environ Policy Law Pap 15:363–389

Durrell G (1977) Zoo breeding. Wildlife 19:53

Edmond-Blanc F (1982) Conférence mondial pour la protection des oiseaux. Saint-Hubert, oct, p 537

Egbert A (1980) Eine Hoffnungsbotschaft. Umweltschutz 7–8:3–27

Ehrlich P (1983) North America after the war. Nat Hist 93:4–8

Ehrlich PR (1980) The strategy of conservation, 1980–2000. In: Soule ME, Wilcox BA (eds) Conservation biology. Sinauer, Sunderland, pp 329–344

Ehrlich PR (1983) Long-term biological consequences of nuclear war. Science 222:1283–1300

Ehrlich PR, Ehrlich AH, Holdren JP (1977) Ecoscience: population, resources, environment. Freeman, San Francisco

Elkiey T, Ormrod D (1979) Ozone and/or sulphur dioxide effects on tissue permeability of petunia leaves. Atmos Environ 13:1165–1168

Elsner R (1983) Die Bedeutung der Änderung der technischen Anleitung zur Reinhaltung der Luft für den Schutz von Pflanzen und Tieren. Natur Recht 5:223–227

Ember RL (1982) Acid rains implicated in forest dieback. Chem Eng News 60:25–26

Ertel R (1982) Weltkonferenz des Internationalen Rates für Vogelschutz. Wir Vögel 14:18–19

Erz W (1976) Über Veränderungen der Brutvogelfauna in der Bundesrepublik Deutschland. Schriftenr Vegetationskd 10:255–267

Everts JW, Frankenhuizen K van, Roman B, Koeman JH (1983) Side-effects of experimental pyrethroid applications for the control of tsetseflies in a riverine forest habitat (Africa). Arch Environ Contam Toxicol 12:91–97

Faber HV (1978) Das Ende vieler Vogelarten. Mitt Dtsch Forschungsgemeinsch 2:21–22

Faber M (1982) Ultraviolet radiation. In: Nonionizing radiat prot. Copenhagen, pp 9–38

Falinski JB (1975) Anthropogenic changes of the vegetation of Poland. Phytocoenosis 4:97–116

FAO (1976a) Development and forest resources in Asia and the Far East: trends and perspectives 1961–1991. Rome, FAO

FAO (1976b) Fisheries technical paper 162

FAO (1977a) Fisheries technical paper 172

FAO (1977b) The state of food and agriculture. FAO, Rome

FAO Forestry Department (1985) Long-range air pollution: a threat to European forests. Unasylva 37:14–25

FAO yearbook of forest products (1977) FAO, Rome

FAO/WHO/UNEP (1976) Guidelines for developing and effective national food control system. FAO, Rome

Farrington JA (1976) Paraquat, peroxide and superoxide. Weeds 1:225

Farthmann F (1983) Luftverunreinigungen und Waldsterben. Forst- Holzwirt 38:49–52

FAS (1977) Federation of American scientists' public interest report. Special Issue: Anim Rights 30:1–8

Fedorenko NP, Reymers NF (1978) Nature and economics. In: Problems of optimization in ecology. Nauka, Moscow, pp 23–39 (in Russian)

Fedorenko NP, Lemeshev MY, Reimers NV (1980) Socio-economic efficiency of conservation. Priroda (Mosc) 10:2–13 (in Russian)

Feldmann R (1984) Gefährdungsursachen und Säugetierschutz. Abh Westfäl Mus Naturkd 46:23–26

Finke L (1980) Ecology and environmental problems. Bochum Geogr Arb 38:54–60

Fischer AC, Peterson FM (1976) The environment in economics: a survey. J Econ Literature 19:1–33

Fleischer CA (1980) The international concern for the environment: the concept of common heritage. IUCN Environ Policy Law Pap 15:321–339, 341–342

Florence TM (1983) Trace element speciation and aquatic toxicology. Trends Anal Chem 2:162–166

Florgärd C (1979) Natur i bostadsomraden? Stadsbyggnad 45:145–150

Flym KC (1982) New challenges in the Great Lake states to banning phosphorus in detergents. Water Pollut Contr Fed 54:1342–1345

Föstner U (1984) Cadmium in sediments. Experientia (Basel) 40:23–29

Fog J (1978) Schutzgebiete für Vögel und Säugetiere in Dänemark. Biol Abh 5:55–58, 56–57

Francis A (1986) Acid rain effects on soil and aquatic microbial processes. Experientia (Basel) 42:455–588

Frankel OH, Soule ME (1981) Conservation and evolution. Cambridge Univ Press, Cambridge, p 327

French JRJ (1980) Australian forest policy. A critical view. Curr Aff Bull 57:4–16

Fuhrer J, Erismann HK (1980) Tolerance of *Aesculus hippocastanum* L. to foliar accumulation of chloride affected by air pollution. Environ Pollut A21:249–254

Fukuoka Y (1982) Dendroclimatological study on urban atmospheric environment. Jpn Progr Climatol 1981, pp 63–71

Fukuto T (1984) Propesticides. In: Magee P, Kohn G, Menn J (eds) Pesticide synthesis through rational approaches. American Chemical Society Washington, pp 87–102

Fyfe R (1976) Bringing back the peregrine falcon. Nature Can 5:10–17

Fyfe R, Ambruster HJ (1977) Raptor research and management in Canada. World Conf Bird of pray, Vienna, Oct 1975. Rept Proc Basinsstoke, pp 282–293

Gaigher IG, Hamman KCD (1980) The distribution, conservation status and factors affecting the survival of indigenous freshwater fishes in Cape province. Koedae 23:57–88

Gaining ground in Galapagos (1978) IUCN Bull, New Ser 9:27

Galiano E, Sterling A, Viejo J (1985) The role of riparian forests in the conservation of butterflies in a Mediterranean area. Environ Conserv 12:361–362

Garrett T (1978) A perspective on our nation's efforts to prevent extinction of species. Inf Rep Anim Welf Inst 27:1–2

Geelen J, Leuven R (1986) Impact of acidification on phytoplankton and zooplankton communities. Experientia (Basel) 42:486–494

Gentile J, Gentile G, Plewa M (1986) In vitro activation of chemicals by plants: a comparison of techniques. Mutat Res 164:53–58

Gerlach Sa (1981) Marine pollution. Springer, Berlin Heidelberg New York, p 218

Gerschenzon SM, Alexandrov Y, Maljuta SS (1975) Mutagenic action of DNA and viruses in *Drosophila*. Naukova Dumka, Kiev

Geyer H, Scheunert I, Korte F (1985) Relationship between the lipid content of fish and their bioconcentration potential of 1,2,4-trichlorobenzene. Chemosphere 14:545–555

Giger W, Brunner P, Schaffner C (1984) 4-Nonylphenol in sewage sluge: accumulation of toxic metabolites from nonionic surfactants. Science 225:623–625

Gilbert LE (1980) Food web organization and conservation of neotropical diversity. In: Soule ME, Wilcox BA (eds) Conservation biology. Sinauer, Sunderland, pp 11–34

The Global 2000. Report to the President. The technical report. (1980) vol 2, Council of environm quality. Washington, p 751

Gissy M (1985) Europe's salt marshes: urgent need for protective network of reserves. Environ Conserv 12:371–372

Global future: time to act. Report to the President on global resources, environment and population (1981). US Council of environm quality, Department of state, p 242

Gold LS, Sawyer CB, Magaw R et al. (1984) A carcinogenic potency database of the standardized of animal bioassays. Environ Health Perspect 58:9–319

Goldberg ED (1975) Synthetic organochlorides in the sea. Proc R Soc Lond B Biol Sci 189:277

Goldberg ED (1976) The health of the oceans. UNESCO, Paris, p 172

Golovleva LA, Skrjabin GK (1979) Co-metabolism of xenobiotics by microorganisms. In: Malakhov SG, Borzilov VA (eds) Migrations and transformations of pesticides in the environment. Gidrometeoizdat, Moscow, pp 40–46

Goodwin WJ (1975) Current status of primate breeding in the United States. Exp Anim 8:221–225

Goryshina TK (1982) Influence of urban air pollution on the photosynthetic apparatus of three tree species. In: Interaction between forest ecosystems and pollutants. Joint Soviet-American symposium. Leningrad-Tallin-Puschchino, October 11–20, 1982. Tallin, pp 86–87

Goryunova SV, Ostroumov SA (1986) Effect of anionic detergent on green algae and seedlings of higher plants. Biol Sci 7:84–86 (in Russian)

Goyal SM (1984) Viral pollution of the marine environment. CRC Critical Rev Environ Control 14:1–39

Graham DS (1983) Lead – the problem is not only in petrol. Gen Eng Trans Inst Eng Aust 7:35–39

Grahn O (1986) Vegetation structure and primary production in acidified lakes in southwestern Sweden. Experientia (Basel) 42:465–470

Grainge M, Ahmed S (1988) Handbook of plants with pest control properties. Wiley, New York, 470 pp

Grebenshchikov OS (1965) Geobotanic dictionary. Nauka, Moscow, p 227

Greene O, Percival I, Ridge I (1985) Nuclear winter: the evidence and the risks. Polity, Cambridge

Greenwood N, Edwards JMB (1973) Human environment and natural systems. A conflict of dominion. Wardswotch, Belmont, p 429

Gregory MR (1983) Virgin plastic granules on some beaches of eastern Canada and Bermuda. Mar Environ Res 10:73–92

Grier JW (1982) Ban of DDT and subsequent recovering of reproduction in bald eagles. Science 218:1232–1235

Grumbach K, Bach T (1979) The effect of PSII herbicides, amitrol and SAN 6706 on the activity of 3-hydroxy-3-methylglutaryl-coenzyme-A-reductase and the incorporation of acetate and mevalonate into chloroplast pigments of radish seedlings. Z Naturforsch C 34:941–943

Grushko YM, Timofeeva SS (1983) Dyes and their detrimental action on organism. Gigiena Sanitarija 8:75 (in Russian)

Guderian R (1977) Air pollution. Springer, Berlin Heidelberg NY, p 127

Hagmann J (1982) Inhibition of calmodulin-stimulated cyclic nucleotide phosphodiesterase by the insecticide DDT. FEBS Lett 143:52–54

Haigh JC, Prorke JS, Parkinson DA, Azcher AL (1979) An elephant extermination. Environ Conserv 6:305–310

Hannah JB, Hose JE, Landolt LM, Miller BS, Felton SP, Twaoka WT (1982) Benzo(a)pyrene-induced morphologic and developmental abnormalities in rainbow trout. Arch Environ Contam Toxicol 11:727–734

Harborne JB (1982) Introduction to ecological biochemistry. Academic Press, London, p 278

Haseltine SD, Finley MT, Cromartie E (1980) Reproduction and residue accumulation in black ducks fed toxaphene. Arch Environ Contam Toxicol 9:461–471

Hatch G, Mamay P, Ayer M, Casto B, Nesnov S (1983) Chemical enhancement of viral transform-ation in Syrian hamster embryo cells by gaseous and volatile chlorinated methanes and ethanes. Cancer Res 43:1954–1950

Hawkesworth DL, Rose F (1970) Quantitative scale for estimating sulphur dioxide air pollution in England and Wales using epiphytic lichens. Nature 227:145–148

Health (1972) Big Soviet Encyclopedia 9:442

Heath J (1981) Threatened Rhopalocera (butterflies) in Europe. Counc Europe Nat Environ Ser 23:157

Heath M (1985) Deep digging for nuclear waste disposal. New Sci 1480:30–32

Hedgpeth JW (1980) The problem of introduced species in management and mitigation. Helgol Meeresunters 33:662–673

Heinemann G (1981) Aktion zur Vermeidung von Wildunfälten in Nordrhein-Westfalen. Wild Hund 83:1168, 1170

Heinisch E, Paucke H, Nagel HD, Hansen D (1976) Agrochemikalien in der Umwelt. Gustav Fischer, Jena

Helander B, Olsson M, Reutergardh L (1982) Residue levels of organochlorine and mercury compounds in unhatched eggs and the relationships to breeding success in white-tailed eagles Haliaeetus albicilla in Sweden. Holarct Ecol 5:349–366

Hemminki K (1981) Teratogenicity testing in experimental animals: an overview. In: Reports of the Second Finnish-Estonian symposium on effects of toxic substances, Helsinki, pp 58–63

Hendrey G (1984) Early biotic responses to advancing lake acidification. Butterworth, Boston, London, Sydney, p 173

Henny CJ (1977) Research, management, and status of the osprey in North America. World Conf Birds of Prey, Vienna, Oct. 1975. Rept Proc Basingstoke, pp 199–222

Henson J (1976) The scope and management of farm parks. Int Zoo Yearb 16:253–254

Herre W, Röhrs M (1971) Über die Verwilderung von Haustieren. Milu 3:131–160

Heseltine M, Peterson R (1982) State of the environment, 1982. International Public Hearing, London, 15–16 June 1982. Environmentalist 2:247–254

Heslop-Harrison J, Lucas G (1978) Plant genetic resources conservation and ecosystem rehabilitation. In: Breakdown and restoration of the ecosystem. Proc Conf Rehabil Severely Damaged and Freshwater Ecosyst Temper Zones, Reykjavik, 1976. New York London, pp 297–306

Hijazi A, Chefurka W (1982) Use of the fluorescent probe, 1-anilino-8-naphtalene sulfonate to monitor the interaction of pesticide chemicals with mitochondrial membranes. Comp Biochem Physiol 73C:369–375

Hileman B (1983) 1982 Stockholm Conference on acidification of the environment. Environ Sci Technol 17:15A–18A

Hilemann B (1982) Radiofrequency and microwave radiation. Environ Sci Technol 16:442A–444A

Hillman K, Martin E (1979) Will poaching exterminate Kenya's rhinos? Oryx 15:131–132

Hoffmann GR (1982) Mutagenicity testing in environmental toxicology. Environ Sci Technol 16:560–574

Hohtola E (1978) Differential changes in bird community structure with urbanisation: a study in Central Finland. Ornis Scand 9:94–100

Holden C (1979) Experts gather to talk turtle. Science 26:1383–1384

Holden C (1984) Zoos forging new role in science. Science 225:147

Holgate MW (1979a) A perspective of environmental pollution. Univ Press, Cambridge, p 278

Holgate MW (1979b) Targets of pollutants in the atmosphere. Philos Trans R Soc Lond A Math Phys Sci 290:591–605

Holling CS (1973) Resilience and stability of ecological systems. Annu Rev Ecol Syst 4:1–23

Hollingshaus JG (1987) Inhibition of mitochondrial electron transport by hydramethylnon: a new amidinohydrazone insecticide. Pestic Biochem Physiol 27:61–70

Honegger RE (1980–1981) List of amphibians and reptiles either known or thought to have become extinct since 1600. Biol Conserv 19:141–158

Hopson JL (1976) Zoos: changing their spots. Sci News 110:106–108

Horio F, Kimura M, Yoshida A (1983) Effect of several xenobiotics on the activities of enzymes affecting ascorbic acid synthesis in rats. J Nutr Sci Vitaminol 29:233–247

Horton FH, Reynolds DR (1971) Effects of urban spatial structure on individual behaviour. Econ Geogr 47:10–15

Howard P, Banerjee S (1984) Interpreting results from biodegradability tests of chemicals in water and soil. Environ Toxicol Chem 3:551–562

Howard PC, Heflich RH, Evans FE, Beland FA (1983) Formation of DNA adducts in vitro and in Salmonella typhimurium upon metabolic reduction of the environmental mutagen 1-nitropyrene. Cancer Res 43:2052–2058

Hudec K (1978) Der Vogelbestand in der städtischen Umwelt von Brno (ČSSR) und seine Veränderungen. Přirodověd pr: Ustavu CSAV Brně 10:1–54

Hutzinger O (ed) (1982) The handbook of environmental chemistry, vol 3, part B. Springer, Berlin Heidelberg New York, p 210

Imaida K, Oshima M, Fukushima S, Ito N, Hotta K (1983) Membrane potentials of urinary bladder epithelium in F344 rats treated with N-butyl-N-(4-hydroxybutyl)nitrosamine. Carcinogenesis 4:659–661

Ingelög T (1985) Sweden: plant update. Threatened Plants Newsletter 15:18–19

Inskipp T, Wells S (1979) International trade in wildlife. Publ Int Inst Environ Dev, London, p 104

Isaev AS, Khlebopros RG, Nedorezov LV, Kondrakov YP, Kiselev VV (1984) Population dynamics of forest insects. Nauka, Novosibirsk, p 224 (in Russian)

Isakov JA, Kazanskaya NS, Panphilov DV (1980) Classification, geography and anthropogenic transformation of the ecosystems. Nauka, Moscow, p 227

IUCN (1985) 1985 United Nations List of National Parks and Protected areas. IUCN. Gland and Cambridge, p 174

IUCN Red Data Book, vol 3. Amphibia and Reptilia (1979)

Ivie GW, Bandal SK (1981) Metabolic aspects of pesticide toxicology. In: Bandal SK, Marco G, Golberg L, Leng M (eds) The pesticide chemist and modern toxicology. Am Chem Soc Wash, pp 257–284

Jackson P (1985) Chancen für den Tiger. Umschau 85:303–308

Jalees K (1985) Pesticides poisoning in India. Chem Ind (Lond) 5:142–143

Jarvinen AW, Nordling BR, Henry ME (1983) Chronic toxicity of Dursban (Chlorpyrifos) to the fathead minnow (Pimephales promenales) and the resultant acetylcholinesterase inhibition. Ecotoxicol Environ Saf 7:423–434

Jaworowski Z (1982) Natural and artificial radionuclides in the Earth's atmosphere. IAEA (International Atomic Energy Agency) Bull 24:38–42

Jellife EEP, Jellife DB (1977) Not safe to fool Mother Nature. Sci News 112:312

Jensen NJ, Block B, Larsen JL (1979) The ulcus-syndrome in cod (Gadus morrhua). A preliminary Virolog. Rep Nord Vet-Med 31:463

Jobst J (1982) Regenschirme für unsere Bäume? Umweltmagazin 6:17–20

Johansen P (1981) Heavy metals in marine mammals and heavy metal intake in humans in Greenland. In: Circumpolar Health 81. Proc Int Symp Copenhagen, 9–13 Aug 1981, Copenhagen, pp 540–542

Johansen P, Kapel FO, Kraul I (1980) Heavy metals and organochlorines in marine mammals from Greenland. ICES C M 1980/E:32

Jones T, Mansfield TA (1982) The effects of SO_2 on growth and development of seedlings of *Phleum pratense* under different light and temperature environments. Environ Pollut (ser A) 27:57–71

Jonston R (1976) Marine pollution. Academic Press, London, p 729

Josephson J (1983) Chlorinated dioxins and furans in the environment. Environ Sci Technol 17:124A–128A

Jule W, Landsdown R, Millar J, Urbanowicz MA (1982) Higher lead levels mean lower IQs for English children. Ambio 11:322–323

Jungius H, Price MS (1980) Present status and future conservation of the Arabian oryx. M.S. IUCN-SSP, pp 1–4

Junqueira V, Simizu K, Videla L, Barros S (1986) Dose-dependent study of the effects of acute lindane administration on rat liver superoxide anion production, antioxidant enzyme activities and lipid peroxidation. Toxicology 41:193–204

Jurgens R (1980) Lurch- und Amphibiensschutz. Naturschutz Naturparke 98:45–46

Kardel L (1979) Japans nationalparker-varldens storsta turistindustri. Fauna Flora 74:155–172

Kaule G (1985) Anforderungen an Grösse und Verteilung ökologischer Zellen in der Agrarlandschaft. Z Kulturtechn Flurbereinig 26:202–207

Kawa K (1979) Zinc-dependent action potentials in giant neurons of the snail, Euhadra quaestia. J Membr Biol 49:325–344

Keighery GJ (1980) Bird pollination in South Western Australia: a checklist. Plant Syst Evol 135:171–176

Kennedy CR (1978) Ecological animal parasitology. Blackwell Sci, Oxford

Kennedy W, Cadwell L, McKenzie D (1985) Biotic transport of radionuclides from a low-level radioactive waste site. Health Phys 49:11–24

Keymer IF (1972) The unsuitability of non-domesticated animals as pets. Vet Rec 91:373–381

Khachaturov TS (1980) Aspects of ecology connected with economics. Vestnik Acad Sci USSR 2:46–52

Khan MAR (1982) Chelonians of Bangladesh and their conservation. J Bombay Nat Hist Soc 79:110–116

Kier L, Brusick D, Auletta A et al. (1986) The *Salmonella typhimurium*/ mammalian microsomal assay. A report of the US EPA GeneTox Program. Mutat Res 168:69–240

King W (1981) The world's rarest birds. Int Wildl 11:12–19

Kinne O, Bulnhaim HP (eds) (1980) Protection of life in the sea. 14th Europ marine biology symposium, Hamburg, p 772

Klekowski EJ, Klekowski EJ (1982) Mutation in ferns growing in an environment contaminated with polychlorinated biphenyls. Am J Bot 69:721–727

Klimek J, Schaap AP, Kimura T (1983) Effect of paraquat on cytochrom P-450-dependent lipid peroxidation in bovine adrenal cortex mitochondria. Biochim Biophys Acta 752:127–136

Klinda J (1980) Ochrana prirody v pol'skej l'udovej republike. Pamiatky Prir 5:16–19

Klingel H (1979) Survey of African rhinoceroses. SSC IUCN, 10.I.80, pp 1–9

Klopatek JM, Olsou RJ, Emerson CJ, Joness JL (1979) Land-use conflicts with natural vegetation in the United States. Environ Conserv 6:191–199

Koivussari J, Nuuja I, Palokangas R, Finnlund M (1980) Relationships between productivity, eggshell thickness and pollutant contents of addled eggs in the population of white-tailed eagles *Haliaëtus albicilla* L. in Finland during 1969–1978. Environ Pollut A23:41–52

Kolb Meyers V (1988) Teratogens. Elsevier, Amsterdam, p 472

Kolesnikov BP, Semenova-Tyan-Shanskaya AM, Dyrenkov SA (1977) Conservation of plants at XII International botanical Congress (Leningrad, 3–10 July 1975). Bot Zh SSSR 62:1792–1809 (in Russian)

Kolman A, Näslund M, Calleman C (1986) Genotoxic effects of ethylene oxide and their relevance to human cancer. Carcinogenesis 7:1245–1250

König C (1981) Schadstoffe, die für Vögel und Fledermäuse den Tod bedeuten. Wir Vögel 13:26

Köpp H (1979) Europäische Ministerkonferenz fur Umweltschutz: Naturschutzkonvention in Bern unterzeichnet. Unser Wald 31:182–183

Kopp S, Perry HM, Perry E, Erlanger M (1983) Cardiac physiologic and tissue metabolic changes following chronic low-level cadmium and cadmium plus lead ingestion in the rat. Toxicol Appl Pharmacol 41:667–680

Koubantsev BS, Zhoukova TI (1982) Some ecological results of man-made effects on populations and habitats of frog *Rana ridibunda*. Ecologiya 6:46–51 (in Russian)

Kozlowski S (1980) Przyspieszmy prawidlowa realizacje programu. Przyr Pol 11:10–11

Kreutzer K, Knorr A, Brosinger F, Kretzschmar P (1983) Scots pinedying within the neighbourhood of an industrial area. In: Eff Accumulat Air Pollutants, Forest Ecosys Proc Workshop, Göttingen, May 16–18, 1982. Dordrecht ea, pp 343–357

Krivolutsky DA (1985) Radioecology of soil animals. Nauka, Moscow, p 209 (in Russian)

Kurinny AI (1985) Bioindication of pesticides-mutagens in the environment. In: Fourth International Conference in Environmental Mutagens. Abstracts, Stockholm, June 24–28, p 389

Kürten W (1980) Landschaftsschutz und Landschaftspflege. Geogr Rdsch 32:194–198, 200–202

Lahti S (1977) Den finska bäverns förekomsthistoria och nutida utbredning. Rapp Och Uppsats Inst Skogszool Skogshögsk 26:19–21

Lamb R (1980) A conference to save the slobs, bogs, mires, marshes, lakes, mudflots, mangroves, saltings of the world. IUCN Bull, New Ser 11:98–100

Landolt E, Fuchs H, Heitz C, Sutter R (1982) Bericht über die gefärdeten und seltenen Gefässpflanzen der Schweiz. Ber Geobot Inst ETH 49:195–218

Landrigan PJ (1982) Occupational and community exposures to toxic metals: lead, cadmium, mercury and arsenic. West J Med 137:531–539

Lang EM (1972) Zoo und Naturschutz. Acta Trop 29:474–481

Laskorin BN (1975) Creation of technologies that exclude detrimental effects of industry on the biosphere, vol 5. Vodnye Resursy (in Russian)

Laskorin BN (1978) Ecology and technology. Chelovek Priroda 10:3–7 (in Russian)

Laughlin R, Linden O (1985) Fate and effects of organotin compounds. Ambio 14:88–94

Laurin H (1983) Sociétés de classification. Protéger les navires c'est aussi protéger la mer. Bull Tech Bur Veritas 65:542–545

Law R, Mellors A, Hallett FR (1985) Physical aspects of the inhibition of enzymes by hydrocarbons: the inhibition of α-chymotrypsin by chlorinated aromatics and alkanes. Environ Res 36:193–205

Leahey JP (ed) (1985) The pyrethroid insecticides. Taylor & Francis, London Philadelphia, 440 pp

Leigh J, Boden R, Briggs J (1984) Extinct and endangered plants of Australia. MacMillan, South Melbourne, 369 pp

Lekyavičius RK (1983) Chemical mutagenesis and environmental pollution. Mokslas, Vilnius, 224 pp

Lelek A (1980) Threatened freshwater fishes of Europe. Counc Europe Nat Environ Ser, Strasbourg 18:269

Lemeshev MY (1978) Welfare of society and environment. Prirodnye Resursy Okruzhayushchaya Sreda 5:17–28 (in Russian)

Leo A (1984) Partitioning in pesticide mode of action and environmental problems. In: Magee P, Kohn G, Menn J (eds) Pesticide synthesis through rational approaches. Am Chem Soc Wash, pp 213–224

Leuven R, Hartog C den, Christiaans M, Heijligers W (1986) Effects of water acidification on the distribution pattern and the reproductive success of amphibians. Experientia (Basel) 42:495–499

Levy EM (1980) Oil pollution and seabirds: Atlantic-Canada 1976–1977 and some implications for Northern environments. Mar Pollut Bull 11:51

Lieth H, Whittaker RH (1975) Primary productivity of the biosphere. Springer, Berlin Heidelberg New York

Linden O (1975) Acute effects of oil and oil/dispersant mixture on larvae of Baltic herring. Ambio 4:126

Lingeman CH (1982) Human resources — measurement of environmental contaminants in post mortem sampling. J Environ Sci Health A A17:515–518

Lipnick R (1985) A perspective on quantitative structure-activity relationships in ecotoxicology. Environ Toxicol Chem 4:255–257

Livingston JA (1980) On the relevance of lungfish, lilacs, wolves, and spirit levels in resource-constrained economies. J Soil Water Conserv 35:165–170

Lloyd OL, Smith G, Lloyd MM, Holland Y, Gailey F (1985) Raised mortality from lung cancer and high sex ratios of births associated with industrial pollution. Brit J Ind Med 42:475–480

Lundborg M, Camner P (1982) Decreased level of lysozyme in rabbit lung lavage fluid after inhalation of low nickel concentrations. Toxicology 22:353–358

Lundqvist LJ (1979) Who is winning the race for clean air? An evaluation of the impacts of the US and Swedish approaches to air pollution. Ambio 8:144–151

Lutz-Ostertag Y, Bruel MT (1981) Action embryotoxique et tératogène du Dichlorvos (insecticide organophosphoré) sur le développement de l'embryon de Caille. CR Acad Sci Paris Sér III 292:1051–1054

Luxmoore RA (1985) Game farming, meat production and conservation. In: Abstr IV ITC, Edmaton, Cauda 13–20, Aug 1985, N 385

Lvov DK, Zhdanov VM (1983) Persistence of genes of epidemical influenza viruses in natural populations in the USSR. Med Biol 61:83–91 (in Russian)

MacCabe RA, Kozicky EL (1972) A position on predator management. J Wildl Manage 36:382–394

MacClenahen JR (1982) Air pollution impacts on forest trees: ecological relationships. In: Interaction between forest ecosystems and pollutants. I Joint Soviet-American symposium. Leningrad, Tallin, Pushchino 11–20 October 1982. Tallin, pp 101–104

MacDonald DW (1980) Rabies and wildlife: a biological perspective. Oxford Univ Press

MacEvoy J, Giger W (1985) Accumulation of linear alkylbenzenesulfonate surfactants in sewage sludges. Naturwissenschaften 72:429–431

MacFarlane BS (1980) Planning for threatened and endangered plant species. J Soil Water Conserv 35:221–223

MacLeod J (1977) And the rivers our blood. NC, Toronto, 112 pp

Maibach HI (ed) (1965) Skin bacteria and their role in infection. MacGraw Hill, New York

Malyshev LI (1981) Changes of floras of Earth under man-made pressure. Biol Nauki (Mosc) 3:5–19 (in Russian)

Mamonov GA (1980) Aquarium hobby and nature conservation. Przegl Zool XXIV:241–247

Marples MJ (1965) The ecology of the human skin. Thomas

Martin RD (ed) (1975) Breeding Endangered species in captivity. Academic Press, London, 1975, p 420

Martynov VA, Novikov RA (1981) Destruction of the environment. Ekonomica, Moscow, p 256 (in Russian)

Mattes H, Eberle C, Schreiber KF (1980) Über den Einfluss von Insektizidespritzungen im Obstbau in die Vitalität und Reproduction von Kohlmeisen (*Parus major*). Vogelwelt 101:132–140

Mattews WH, Smith FE, Goldberg ED (1971) Man's impact on terrestrial and oceanic ecosystems. MIT Press, Cambridge

Maximov VN, Nagel H, Kovaleva TN, Ostroumov SA (1988a) Bio-assay of water polluted by sulfonol. Water Resources 1:165–168

Maximov VN, Nagel H, Ostroumov SA (1986) Experimental study of the effects of water pollution by detergents on *Fagopyrum esculentum* seedlings. In: Izrael YA (ed) Problems of ecological monitoring and ecosystem modelling, vol 9. Gidrometeoizdat, Leningrad, pp 87–97

Maximov VN, Nagel H, Ostroumov SA (1988b) Bio-assay of water containing surfactant and DNOC. Hydrobiological Journal 24:54–55

Mayers M (1982) Saving the African elephant. Zoonooz 55:4–9

Menshikova LN (1983) On gonadotropic effects of electric field of low frequence. In: Second All-Union conference: Endocrine system of the organism and harmful factors of environment. 21–23 September, 1983. Leningrad, p 137 (in Russian)

Merian E (ed) (1984) Metalle in der Umwelt. Chemie Deerfield Beach (Florida) Basel, p 722

Meyers TR, Hendricks JD (1982) A summary of tissue lesions in aquatic animals induced by controlled exposures to environmental contaminants, chemotherapeutic agents and potential carcinogens. Mar Fish Rev 44:1–17

Meyer-Spasche H (1982) Streusalzschäden und Möglichkeiten der Beseitigung. Gartenamt 31:673–379

M'Gonigle RM, Zacher MW (1979) Pollution, politics, and international law. Tanker at sea. Berkley, p 394

Mietz J, Sjogren R (1983) Incidence of plasmid-linked antibiotic-heavy metal resistant enterics in water-sediment from agricultural and harbor sites. Water Air Soil Pollut 20:147–159

Miller DS, Hallett DJ, Peakall DB (1982) Which components of crude oil are toxic to young seabirds? In: Ward CH (ed) Environmental toxicology and chemistry. Pergamon, New York, pp 39–44

Miller G, Connell D (1982) Global production and fluxes of petroleum and recent hydrocarbons. Int J Environ Stud 19:273–280

Miller R, Harris LD (1979) Predicting species changes in isolated wildlife preserves. In: Proc 1st Conf Res Nat Parks, New Orlando, La, 1976, vol 1. Washington DC, pp 79–82

Milliken T (1985) Concern over Japanese bear trade. Traffic Bull VII:5–8

Mironov OG (1985) Interaction of marine organisms with oil hydrocarbons. Hydrometeoizdat, Leningrad, p 128 (in Russian)

Mitchell D (1982) First extinct Arabian oryx born in the wild. Environ Conserv 9:58

Miyamoto J, Kaneko H, Hutson D, Esser H, Gorbach S, Dorn E (1988) Pesticide metabolism: extrapolation from animals to man. Blackwell Oxford 120 p

Mizgireuv IV, Flax NL, Borkin LJ, Khudolev VV (1984) Dusplastic lesions and abnormalities in amphibians associated with environmental conditions. Neoplasma 31:175–181

Moiseev NN, Alexandrov VV, Tarko AM (1985) Man and biosphere. Nauka, Moscow, p 271 (in Russian)

Monaghan P, Coulson JC (1977) Status of large gulls nesting on buildings. Bird Study 24:89–104

Monitor (1985) The national Swedish Environmental monitoring programme (PMK). Stockholm, p 209

Moore J (1985) Science as way of knowing — human ecology. Am Zool 25:483–637

Moore N (1982) Lost British dragonflies. Oryx 16:316–317

Moore P (1985) The ecology of diversity. New Sci 105:17–19

Mostafa IY, Adam YM, Zayed SM (1983) Bioalkylation of nucleic acids in mice by insecticides. I. Alkylation of liver RNA and DNA by chlorpyrifos. Z Naturforsch 38c:461–464

Mudd JB, MacManus TT, Ongun A, MacCullogh TE (1971) Inhibition of glycolipid biosynthesis in chloroplasts by ozone and sulfhydryl reagents. Plant Physiol 48:335–339

Mudrack K, Kunst S (1988) Biologie der Abwasserreinigung. Fischer, Stuttgart 178 p

Mukherji S, Mukerji C (1981) Role of cadmium in regulating pigment efflux from beet (*Beta vulgaris* L.) root tissues. Indian J Exp Biol 19:197–198

Mulder G (1984) The role of sulfation and glucuronidation in toxification of xenobiotics. In: Caldwell J, Paulson G (eds) Foreign compound metabolism. Taylor & Francis, London, pp 235–244

Murphy JR (1977) Eagles and livestock-save management considerations. World Conf Birds of Prey, Vienna, Oct. 1975. Rept Proc Basingstoke, pp 307–314

Murty AS, Rajabhushanam BR, Ramani AV, Christopher K (1983) Toxicity of fenitrothion to the fish *Mystus cavasius* and *Labeo rohita*. Environ Pollut A 30:225–232

Muslih RK, Linscott DL (1977) Regulation of lipid synthesis in soybeans by two benzoic acid herbicides. Plant Physiol 60:730–735

Myers N (1980a) Conversion of tropical moist forests. National Academy of Sciences, Washington, p 205

Myers N (1980b) The sinking ark. A new look at the problem of disappearing species. Pergamon, Oxford, pp 307

Myers N (1980c) The problem of disappearing species: what can we done? Ambio 9:229–235

Myers N (1980d) The present status and future prospects of tropical moist forests. Environ Conserv 7:101–114

Myers N (1983a) A wealth of wild species: storehouse for human welfare. Westview, Boulder, 274 pp

Myers N (1983b) The tropical forest issue. In: O'Riordan T, Turner RK (eds) Progress in resource management and environmental planning, vol 4. Wiley, pp 1–28

Myers N (1984) How cats have run out of lives. Futures, 8 November

Myers N (1988) Tropical forests and their species going, going . . . In: Wilson EO (ed) Biodiversity. National Academy Press, Washington, pp 28–35

Nakachi K, Kubo T, Sasaba T, Yonamine M (1981) The geographical distribution of cancer deaths in Saitama prefecture, Japan, studied by modified principal component analysis. Med Biol Environ 9:28–36

Nakada S, Nomoto A, Onozaki K, Imura N (1981) Methyl mercury inhibits lectin-mediated cell agglutination. Ecotoxicol Environ Saf 5:437–442

Narahashi T (1982) Modification of nerve membrane sodium channels by the insecticide pyrethroids. Comp Biochem Physiol C Comp Pharmacol 72:411–414

National Res Council (1977) World food and N nutrition study: supporting papers, vol 1. Nat Acad Sci, Wash

Negilski DS, Ahsanullah M, Mobley BC (1981) Toxicity of zink, cadmium and copper to the shrimp Callianassa australiensis. II. Effect of paired and triad combinations of metals. Mar Biol 64:305–309

Neuberger JS, Hollowell JG (1982) Lung cancer excess in an abandoned lead-zink mining and smelting area. Sci Total Environ 25:287–294

Neumann W, Laasch H, Urbach W (1987) Mechanisms of herbicide sorption in microalgae and the influence of environmental factors. Pestic Biochem Physiol 27:189–200

Ney P (1986) Asbestos. In: Hutzinger O (ed) The handbook of environmental chemistry, vol 3 D. Springer, Berlin Heidelberg New York, pp 35–100

Nichols FH, Cloern JE, Luoma SN, Peterson DH (1986) The modification of an estuary. Science 231:567–573

Nikitin DP, Novikov VV (1980) Environment and man. Vysshaya Shkola, Moscow, p 424 (in Russian)

Nikolaev II (1980) Towards a theory of ecological prognostication of lake ecosystems. Water Res 5:100–109 (in Russian)

Nilsson G (1983) The endangered species handbook. Animal Welfare, Inst, Washington, p 245

Nitecki C (1980) Organizacja, zadania organów oraz środki prawne w ochronie naturalnego środowiska w Polsce. Przem Drzew 31:18–20

Van Note G (1978) Endangered species legislation. Science 201:1174

Nriagu J (ed) (1979) Copper in the environment, part 1. Wiley, New York

Nyberg H, Koskimies-Soininen K (1984) The phospholipid fatty acids of *Porphyridium purpureum* cultured in the presence of Triton X-100 and sodium desoxycholate. Phytochemistry 23:2489–2495

Obzor fonovogo sostojanija okruzhajushchey prirodnoy sredy v SSSR za 1988 (Survey of the state of environment in the USSR in 1988) (1989) Gidrometeoizdat Press, Moscow, 102 pp

Odum EP (1971) Fundamentals of ecology, 3rd edn. Saunders, Philadelphia, p 742

OECD (1979) The state of the environment in OECD member countries. Organ for economic co-operation and development, Paris

Oettmeier W, Kude C, Soll HJ (1987) Phenolic herbicides and their metylethers: binding characteristics and inhibition of photosynthetic electron transport. Pestic Biochem Physiol 27:50–60

Økland J, Økland K (1986) The effects of acid deposition on benthic animals in lakes and streams. Experientia (Basel) 42:471–486

Oldfield ML (1984) The value of consering genetic resources. US Dept of the interior, Nat park Service, Washington, p 360

Oliver B, Niimi A (1983) Bioconcentration of chlorobenzenes from water by rainbow trout: correlations with partition coefficients and environmental residues. Environ Sci Technol 17:287–291

Olney PJS (ed) (1984/1985) International Zoo Yearbook. vol 24/25. Zoological Society of London, London p 651

Olney PJS et al. (eds) (1981) 1981 International Zoo Yearbook, vol 21. Zool Soc, London, p 404

Olsson M (1977) Mercury, DDT and PCB in aquatic test organisms. Nat Swedish Environ Protect Board, Stockholm p 199

Opler PA (1977) The parade of passing species: a survey of extinction in the USA. Sci Teacher 44:1–6

Ostro B (1984) A search for a threshold in the relationship of air pollution to mortality: a reanalysis of data on London winters. Environ Health Perspect 58:397–399

Ostroumov SA (1984) Biochemistry and environmental protection: in search of regulators. Znanie, Moscow, p 69 (in Russian)

Ostroumov SA (1986) Introduction to biochemical ecology. Moscow Univer, Moscow, p 176

Ostroumov SA (1990) Assessment of biological activity of xenobiotics. Vestnik Moscovskogo universiteta (Bulletin of Moscow University, in Russian), Biological Series (in press)

Ostroumov SA, Jasaitis AA, Samuilov VD (1979) Electrochemical proton gradient across the membranes of photophosphorylating bacteria. In: Manson LA (ed) Biomembranes. Plenum Press, NY London, vol 10, pp 209–234

Ostroumov SA, Kaplan A Ya, Kovaleva TN, Maximov VN (1988) A study of some aspects of ecotoxicology of sulfonol in plants and other organisms. In: Krivolutzky DA (ed) Ecotoxicology and conservation. Institute of Biology, Riga, pp 134–136

Ostroumov SA, Maximov VN (1988) Disturbance of onthogenesis of Camelina sativa and of Triticum aestivum under effect of non-ionogenic surfactant. In: Krivolutzky DA (ed) Ecotoxicology and conservation. Institute of Biology, Riga, pp 133–134

Ostroumov SA, Tretjakova AN (1990) Effects of environment pollution by cationic surfactant on algae and seedlings of *Fagopyrum esculentum* Moench. Ekologija (Ecology, USSR) (in press)

Ostroumov SA, Vorobiev LN (1976) Membrane potential as a possible polyfunctional regulator of activities of membrane proteins. Biol Nauki (Biological Sciences, in Russian) 7:22–26

Ostroumov SA, Vorobiev LN (1978) Membrane potential and surface charge densities as possible generalized regulators of membrane protein activities. J Theor Biol 75:289–297

Owen-Smith N (1985) Pleistocene extinction: the pivotal role of megaherbivores. Abst IV JTC. Edmonton, Canada, 13–20 Aug. No 477

Paine R (1966) Food web complexity and species diversity. Am Nat 100:65–76

Panwar HS (1982) Tiger ten years later. Ambio 11:330–337

Parker JCS, Martin EB (1979) Trade in African rhino horn. Oryx 15:153–158

Pashin YuV, Kozachenko VI, Zatsepilova TA, Bakhitova LM (1983) Chemical mutagens of environment. Nauka, Moscow, p 139 (in Russian)

Patin SA (1979) Effects of pollution on biological resources and productivity of world ocean. Legpishchprom, Moscow, p 304 (in Russian)

Paucke H (1980) O vyvoji ochrany zivotneho prostredia v NDR. Zivot Prostr 14:103–104

Paulson GD (1984) The economic and toxicological importance of comparative xenobiotic metabolism. In: Caldwell J, Paulson GD (eds) Foreign compound metabolism. Taylor & Francis, London Philadelphia, pp 125–132

Pavan M (1969) SOS planet Earth. Lito-Tipo M Ponzio, Pavia, p 211

Pawar KR, Ghate HV, Katdare M (1983) Effect of malathion on embryonic development of the frog *Microhyla ornata* (Dumeril and Bibron). Bull Environ Contam Toxicol 31:170–176

Peakall DB, Miller DS, Kinter WB (1983) Toxicity of crude oil and their fractions to nestling herring gulls. I. Physiological and biochemical effects. Mar Environ Res 8:63–71

Pearsall MH (1962) In: Dhelren E, Holgate MW (eds) The exploitation of natural animal populations. Brit Ecolog Soc Symp No 2, Oxford, Blackwell

Peattie M, Lindsay D, Hoodless R (1984) Dietary exposure of man to chlorinated benzenes in the U.K. Sci Total Environ 34:73–86

Pelkonen O, Nebert D (1982) Metabolism of polycyclic aromatic hydrocarbons: etiologic role in carcinogenesis. Pharmacol Rev 34:189–216

Perrson R (1974) World forest resources. Royal College of Forestry, Stockholm

Perry J, Bridgwater DD, Horsemen DL (1975) Captive propagation: a progress report. Breed Endangered Spec Captivity, London, pp 361–372

Perry J, Bridgwater DD, Horsemen DL (1982) Captive propagation: a progress report. Zoologica (NY) 57:109–117

Peskov V (1982) Birds on aerial conductors. Molodaya Gvardia press, Moscow, p 304 (in Russian)

Peters R, Darling J (1985) The greenhouse effect and nature reserves. Bioscience 35:707–716

Peterson J (1984) Through the 21st century. Ambio 13:134–141

Petzold HG (1979) Krokodilschutz auf Kuba-beispielhaft für die Welt. Urania 55:18–22

Pianka ER (1966) Latitudinal gradients in species diversity: a review of concepts. Am Nat 100:33–46

Pianka ER (1978) Evolutionary ecology, 2nd edn. Harper & Row, New York, p 400

Picton HD (1979) The application of insular biogeographic theory to the conservation of large mammals in the Northern Rocky Mountains. Biol Conserv 15:73–79

Pierson W, Chang T (1986) Acid rain in western Europe and northeastern United States — a technical appraisal. CRC Crit Rev Environ Control 16:167–192

Pilinskaya MA (1986) Mutagenic effects of pesticides. Review of Science and Engineering, vol 9. General Genetics. Viniti, Moscow, pp 97–151 (in Russian)

Pimentel D, Edwards C (1982) Pesticides and ecosystems. Bioscience 32:595–600

Pimentel D, Levitan L (1986) Pesticides: amounts applied and amount reaching pests. Bioscience 36:96–91

Pimentel D, Pimentel S (1980) Ecological aspects of agricultural policy. Nat Res J 20:555–585

Pinder NJ, Barkham JP (1978) An assessment of the contribution of captive breeding to the conservation of rare mammals. Biol Conserv 13:187–245

Ponomarev V, Kasumjan A, Lukjanov A (1983) Change of defence behavior of fish in response to alarm pheromone under phenol intoxication. In: Sokolov VE, Zinkevich EP (eds) 2nd Conference on chemical communication of animals. Abstracts. Moscow, p 119

Poore D (1978) Tropical rain forests and moist decidous forests (Sourcebook for a World Conservation Strategy) IUCN, p 30

Popova P, Profirov Y (1982) On the permeability of entherocyte membrane under the action of Agria-1060. Anim Breed Sci 19:114–119 (in Bulgarian)

Porter WP, Hindsdill R, Fiarbrother B et al. (1984) Toxicant-disease-environment interaction associated with suppression of immune-system, growth and reproduction. Science 224:1014–1017

Priklonsky SG (1977) Conference on game ornithology. Zool Zh (Zool J) 56:173–174 (in Russian)

Protasov VR (1982) Electro-physical antropogenet fields in the aquatic areas. Vestn AN SSSR 9:71–79 (in Russian)

Ptaszyk J (1979) Negatiwne oddzialywanie chemizacji, mechanizacji komunikalji na ptaki. Chronmy Przyr Ojczysta 35:52–59

Pudlis E (1983) Poland's plight: environment damaged from air pollution and acid rain. Ambio 12:125–127

Rabotnov TA (1983) Phytocoenology. Moscow Univ, Moscow, p 292

Racey DA (1979) Two bats in the Seychelles. Oryx 15:148–152

Raczynski J (1975) Progress in breeding European bison in captivity. In: Martin RD (ed) Breeding endangered species in captivity. Academic Press, London, pp 253–262

Raffin JP (1983) D'une introduction à l'autre. Courr Nat 86:7–10

Raffin JP, Godineay JC, Ribier J, Platel R, Francillon H, Meunier F, Benest G (1981) Evolution sur trois années de quelques sitex littoraux de Bretagne après pollution pétrolière (Amoco Cadiz). Cah Biol Mar 22:323–348

Raikow R, Okunewick J, Jones D, Buffo M (1983) Potentiation of Friend viral leukemogenesis by 9,10-dimethyl-1,2-benzanthacene in two strains of mice. Proc Soc Exp Biol Med 173:125–129

Rajan TPS (1985) The environmental crisis. Chem Weekly 30:1–2

Rall DP, MacLachlan JA (1980) Potential for exposure to estrogens in the environment. In: MacLachlan JA (ed) Estrogens in the environment. Elsevier, North Holland, pp 199–202

Ramade F (1978) Éléments d'écologie appliquée. McGraw-Hill, New York

Randall J (1980) Conserving marine fishes. Oryx 15:287–291

Randall RM, Randall BM, Bevaur J (1980) Oil pollution and pinguins – is clearing justified? Mar Pollut Bull 11:234–237

Ratcliffe R (1983) Children's IQ peril: new lead ban demand. The Sunday Times, 6.03.83, N 8278:1–2

Raven P (1981) Research in botanical gardens. Bot Jahrb Syst 102:53–72

Raven PH (1976) The destruction of the tropics. Frontiers 40:22–23

Ray CG, Dobbin JA, Salm RV (1976) Strategics for protecting marine mammals habitats. Oceanus: 55–67

Red Data Book IUCN (1978) Plants

Red Data Book IUCN (1979) Amphibians

Red Data Book IUCN (1982) Birds

Red Data Book IUCN (1983) Invertebrates

Red Data Book IUCN (1984) Swallowtail butterflies

Red Data Book of the RSFSR (Russian Soviet Socialist Republics, in Russ) (1983) Rosselkhozizdat, Moscow, p 455

Red Data Book of the USSR (1984a) vol 1 (animals) p 392

Red Data Book of the USSR (1984b) vol 2 (plants) 2nd edn. Lesnaya Prom, Moscow, p 480

Regenstein L (1975) The politics of extinction. MacMillan, New York

Regier HA, Cowell EB (1972) Applications of ecosystem theory, succesion, diversity, stability, stress and conservation. Biol Conserv 4:83–88

Reijnders L (1981) Kinderen erstig bedreigd door lood. Natuur Milieu N 11–12:19–22

Reinfeld U (1979) Die Feuchtgebiete Bayerns. Bayerland 81, N 12:13–20

Renwick A (1983) Unmetabolized compounds. In: Caldwell J, Jacoby W (eds) Biological basis of detoxication. Academic Press, London, pp 151–179

Retief GP (1971) The potential of game domestication in Africa, with special reference to Botswana. JS Afr Vet Med Assoc 42:119–127

Reymers NF, Shtilmark FR (1978) Specially protected natural areas. Mysl, Moscow, p 296 (in Russian)

Richardson B, Waid J (1982) Polychlorinated biphenyls. Search (Syd) 13:17–25

Richmond M (1973) Land animals. Ann N Y Acad Sci 216:121–127

Riphagen W (1980) The international concern for the environment as expressed in the concepts of the "common heritage of mankind" and of "shared natural resources". IUCN Environ Policy Law Pap 15:343–362

Rivera CM, Penner B (1979) Effect of herbicides on plant cell membrane lipids. Res Rev 70:45–76

Roberts NJ, Michaelson S, Lu S (1986) The biological effects of radiofrequency radiation: a critical review and recommendations. Int J Radiat Biol 50:379–420

Rochow JJ (1980) Power plant water intake assessment. Environ Sci Technol 14:398–402

Rodin LE, Başilevich NI (1965) Dynamics of the organic matter and biological turnover of ash elements and nitrogen in the main types of the world vegetation. Nauka, Moscow, p 253 (in Russian)

Roelofs J (1986) The effect of airborne sulphur and nitrogen deposition on aquatic and terrestrial heathland vegetation. Experientia (Basel) 42:372

Romer AS (1966) Vertebrate paleontology. The University of Chicago Press. Chicago, London, 468 p

Roonwal ML (1977) Urbanization in primates and comparison with man. J Sci Ind Res 36:179–187

Rose FL, Harshbarger JC (1977) Neoplastic and possibly related skin lesions in neotenic tiger salamandres from a sewage lagoon. Science 196:315–317

Rotmistrov MN, Gvozdjak PI, Stavskaya SS (1978) Microbiology of sewage water treatment. Naukova Dumka, Kiev, p 267 (in Russian)

Ruggiero P, Radogna VM (1985) Inhibition of soil humus-laccase complexes by some phenoxyacetic and s-triazine herbicides. Soil Biol Biochem 17:309–312

Rush GF, Smith JH, Bleavins M, Aulerich RJ, Ringer RK, Hook JB (1983) Perinatal hexachlorobenzene toxicity in the mink. Environ Res 31:116–124

Rychkowski L (ed) (1974) Ecological effects of intensive agriculture (First attempt at a synthesis). Warsawa

Ryder OA (1985) The Asiatic wild horse, *Equus przewalskii* (Poljakov 1881) in captivity. Abst IV ITC, Edmonton, Ca, 13–18 Aug 1985, No 543

Safe S (1985) Polychlorinated biphenyls (PCBs) and polybrominated biphenyls (PBBs): biochemistry, toxicology, and mechanism of action. CRC Crit Rev Toxicol 13:319–352

Safe S (1986) Comparative toxicology and mechanism of action of polychlorinated dibenzo-p-dioxins and dibenzofurans. Annu Rev Pharmacol Toxicol 26:371–400

Sage B (1979) Hawaii — paradise lost? New Sci 84:682–685

Sager P, Doherty R, Olmsted J (1983) Interaction of methylmercury with microtubules in cultured cell in vitro. Exp Cell Res 146:127–138

Saint-Mark P (1971) Socialisation de la nature. Paris, p 434

Sainz-Ollero H, Hernandez-Beunejo IE (1979) Experimental reintroductions of endangered plant species in their natural habitats in Spain. Biol Conserv 16:195–206

Saith W (1982) Air pollution impacts on forest trees. In: Interaction between forest ecosystems and pollutants. I. joint Soviet-American symposium on the project 02:03–21. Leningrad, Tallin, Pushchino 11–20 Oct 1982, Tallin, pp 15–17

Sandala MG, Dumanskii HD, Rudnev MI, Ershova LK, Los IP (1979) Study of nonionizing radiation effects upon the central nervous system and behavior reactions. Environ Health Perspect 30:115–121

Sanders HL, Grassle JF, Hampson GR, Morse LS, Garner-Price S, Jones CC (1980) Anatomy of an oil spill: long-term effects from the grounding of the barge Florida off West Falmouth, Massachusetts. J Mar Res 38:265–380

Sano A, Satoh N, Kubo N (1979) Floating particulate petroleum residues tar balls, in the western Pacific. Oceanogr Mag 30:47-53

Sanotsky IV, Salnikov LS (1978) Problems of embryotropic effects of environmental chemical factors. In: Ecological prognostication. Nauka, Moscow, pp 236-260 (in Russian)

Schefler W (1980) Rabics and wildlife: a perspective. ANS Inf Rep 29:7

Schlee D (1986) Okologische Biochemie. Gustav Fischer Jena, p 355

Schneider T, Grant L (1982) Air pollution by nitrogen oxides. Elsevier, Amsterdam, p 1100

Schultz TW, Applehaus F (1985) Correlation for the acute toxicity of multiple nitrogen substituted aromatic molecules. Ecotoxicol Environ Saf 10:75-85

Schutze C (1979) Heimat fur Flatterkaffe gesucht. Natur Umwelt 59:131-132

Scott AD (1955) Natural resources: the economics of conservation. Univers Toronto Press, Toronto

Seebeck JH (1977) Mammals in the Melbourne metropolan area. Victorian Nat 94:165-171

Sentzova OY, Maximov VN (1985) Action of heavy metals on microorganisms. Adv Microbiol 20:227-252 (in Russian)

Shandala NG, Rudnev MI, Vinogradov GI (1985) Biological effects of low-intensity microwave radiation as an environmental factor. In: State-of-the-art and prospects of environmental hygiene: methods, theory and practice. Nauka, Moscow, pp 173-178 (in Russian)

Shotts E, Vanderwork V, Campbell L (1976) Occurrence of R factors associated with *Aeromonas hydrophila* isolated from aquarium fish and waters. J Fish Res Board Can 33:736-740

Shrinner DS (1982) Air pollution impacts on forest trees: acid deposition. In: Interaction between forest ecosystems and pollutants. I Joint Soviet-American symposium on project 02.03-21, Leningrad, Tallin, Pushchino 11-20 Oct 1982, Tallin, pp 132-134 (in Russian)

Sieberth H (1980) Neue Ansätze in Naturschutz einer Großstadt. Gartenamt 29:344-353

Siljak DD (1975) When is a complex ecosystem stable? Math Biosci 25:25-50

Simberloff D, Gotelli N (1984) Effects of insularisation on plant species richness in the prairie-forest ecotone. Biol Conserv 29:27-46

Simberloff DC, Abele LG (1976) Island biogeography: theory and conservation practice. Science 191:285-286

Simpson R (ed) (1984) Land management for conservation. Eur Conf Castleton, 23-27 Apr 1984. Bakewell, Peak Nat Park Cent, p 91

Sire J (1983) La destruction des animaux nuisibles. Saint-Hubert, pp 114-145

Skärby L, Sellden G (1984) The effects of ozone on crops and forests. Ambio 13:68-72

Skelly JM (1982) Air pollution impacts on forest trees: Growth and productivity. In: Interactions between forest ecosystems and pollutants. I Joint Soviet-American symposium on project 02.03-21, Leningrad, Tallin, Pushchino, 11-20 Oct 1982, Tallin, pp 93-96 (in Russian)

Skiner J (1972) The springbook: a farm animal of the future. Afr Wildl 26:114-115

Sly JMA (1977) Study of man's impact on climate; inadvertent climate modification. MIT Press, Cambridge

Smart NDE, Hatton JC, Spence DHN (1985) The effect of long-term exclusion of large herbivores on vegetation of Murchison Falls national park, Uganda. Biol Conserv 33:229-245

Smirnov MN (1983) Wild animals of South Siberia. Priroda 11:76-83

Smith CJ, Wijngaarden A van (1976) Threatened mammals (council of Europe). Res Int for Nature Management Kasteel Broekhuize zen-Leersum-NL

Smith J, Witkowski P, Fusillo T (1988) Manmade organic compounds in the surface waters of the United States — a review of current understanding. US Geological Survey Circular 1007:1-92

Smith NJH (1979) Aquatic turtles of Amazonia: an endangered resource. Biol Conserv 16:165-176

Smith NJH (1981) Caimans, capybaras, otters, manatees and man in Amazonia. Biol Conserv 19:177-187

Smith W (1981) Air pollution and forests. Springer, Berlin Heidelberg New York, p 379

Söderlund R, Rosswall T (1982) The nitrogen cycles. In: Hutzinger O (ed) The handbook of environmental chemistry, vol 1, part B. Springer, Berlin Heidelberg New York, pp 61-82

Sommer A (1976) Attempt at an assessment of the world's tropical forests. Unasylva 28 (112/113):5-25

Sorensen P (1983) Investigations into the origin(s) of the freshwater attractant(s) of the American eel. In: Muller-Schwarze D, Silverstein RM (eds) Chemical signals in vertebrates, vol 3. Plenum, New York, pp 313-316

Soule ME (1980) Thresholds for survival maintaining fitness and evolutionary potential. In: Soule ME, Volcox BA (eds) Conservation biology. Sinauer, Sunderland, pp 151–170

Soule ME, Wilcox BA (eds) (1980) Conservation biology. Sinauer, Sunderland, p 395

Spectrum (1979) Environment 21:21–24

Spectrum (1980) Environment 22:21–24

Spectrum (1981) Environment 23:21–24

Speight M (1985) European insects. Naturopa 49:4–6

Stachel B, Gabel B, Kozicki R, Lahl U, Podbielsk A, Schlosser M, Zeschmar B (1983) Gibt es einen Zusammenhang swischen der Fruchtbarkeit des Mannes und den Belastungen mit Umwelt-chemikalien? GIT 27, suppl: Umwelt/Sicherh, pp 12–13

Stanley PI, Bunyan PJ (1979) Hazards of wintering geese and other wildlife from the use of dieldrin, chlofenilphos and carbophenothoin as weat treatment. Proc R Soc Lond B Biol Sci 205:31–45

Steiert JS, Crawford RL (1985) Microbial degradation of chlorinated phenols. Trends Biotechnol 3:300–305

Sten L (1977) Baverns utbredningshistoria och nuvarande forekomst i Sverige. Rapp Och Uppsats Inst Skogzool Skoghogsk 26:8–12

Stohs S, Hassan M, Murray W (1983) Lipid peroxidation as a possible cause of TCDD toxicity. Biochem Biophys Res Commun 111:854–859

Stugren B (1986) Grundlagen der allgemeinen Okologie. Gustav Fischer. Jena, p 356

Sumerling T, Dodd N, Green N (1984) The transfer of strontium-90 and caesium-137 to milk in a dairy herd grazing near a major nuclear installation. Sci Total Environ 34:57–72

Sun M (1984a) Static at EPA over broadcast transmitters. Science 225:32–33

Sun M (1984b) Use of antibiotics in animal feed challenged. Science 226:144–146

Svein M (1977) Beverens utbredelse i Norge omkring 1975. Rapp Och Uppsats Inst Skogzool Skogshogsk 26:13–18

Svirezhev YM, Alexandrov GA, Arkhipov PL et al. (1985) Ecological and demographic consequences of nuclear war. Computer Center of the USSR. Academy of Sciences, Moscow, p 267

Svoboda FJ (1980) A wildlife planning process for private landowners. Wildl Soc Bull 8:98–104

Szaro RC, Coon NC, Stout W (1980) Weathered petroleum: effects on mallard egg hatchability. J Wildl Manage 44:709–713

Tabacova S (1985) Ambient air pollution in relation to mortality. Nutr Res, Suppl N 1:670–673

Takhtajan A (ed) (1981) Rare and vanishing plants of the USSR. Nauka, Leningrad, p 263

Tamisier A (1981) Zones humides et oiseaux d'eau. Nouv Environ, sa, N 2, p 5

Tanaka T, Ikemura K, Sunoda M, Sasagawa I, Mitsuhashi S (1976) Drug resistance and distribution of R factors in Salmonella strains. Antimicrob Agents Chemother 9:61–64

Taylor PG (1985) Fallow deer farming in Australia. Abstr IV ITC, Edmonton, Canada 13–20 Aug 1985, N 615

Telitchenko MM, Ostroumov SA (1990) Introduction to problems of biochemical ecology. Nauka, Moscow (in press)

Tell RA, Mantyply ED (1982) Population exposure to VHF and UHF broadcast radiation in the United States. Radio Sci 17, N 5, Suppl: Helsinki Symp Biol Eff Electromag Radiat, pp 39–47

Temple SA (1977) Plant-animal mutualism: coevolution with Dodo leads to near extinction of plant. Science 6277:885–886

Terrasse M (1980) Vers une éthique de la réintroduction dans la nature. Homme Oiseau 18:175–179

Terrasse M (1982) Le retour des boulgras. Courr Nat 79:15–24

Thielcke G (1982) Eine Vogel-Bilanz. Natur Umwelt 62, N 4:8–9

Tikhomirov FA (1983) Radioecology of iodine. Energoatomizdat, Moscow, p 88

Tischler W (1965) Agrarokologie. VEB Fischer, Jena

Tokiwa H, Kitamori S, Nakagawa R, Horikawe K, Matamala L (1983) Demonstration of a powerful mutagenic dinitropyrene in airborne particulate matter. Mutat Res 121:107–116

Tomlinson DNS (1980) Nature conservation in Rhodesia: a review. Biol Conserv 18:159–177

Traffic News (1980) CITES uncovers fur trade scandal. IUCN Bull 11:108

Train RE (1979) Back from the dead. The New York Times, July 11:A8–A9

Treshiw M, Anderson FK, Harner F (1967) Responses of Douglas fir to elevated atmospheric fluorides. For Sci 13:114–120

Troitskaya MN, Ermolaeva-Makovskaya AP, Ramzaev PV (1982) A role of radioactivity in carcinogenesis in Arctic population. In: Space study of anthropoecological situation in Siberia and Far East. I All-Union conference on space anthropoecology. Novosibirsk 21-24 December 1982. Leningrad, pp 88-89

Trost RE (1980) Ingested shot in waterfowl harvested on the Upper Mississippi National Wildlife Refuge. Wildl Sci Bull 8, N 1:71-74

Turco RP, Toon DB, Ackerman TP et al. (1983) Nuclear winter: global consequences of multiple nuclear explosions. Science 222:1283-1292

Turco RP, Toon O, Ackerman TP, Pollack J, Sagan C (1984) The climatic effects of nuclear war. Sci Am 251, N 2:33-43

Turtle EE (1963) The effect on birds of certain chlorinated insecticides used as seed dressings. J Sci Food Agric 14:567-577

Tuthill RW, Giusti RA, Moore GS, Calabrese EJ (1982) Health effects among newborn after prenatal exposure to ClO_2-disinfected drinking water. Environ Health Perspect 46:39-45

Underwood L (1983) The farm for butterflies. Anglia (England) 86:61-67 (in Russian)

UNEP (1977) The state of the environment: selected topics — 1977. Pergamon, Oxford

UNEP (1982) The state of the environment. UNEP, Nairobi, p 63

Valenta P, Nguyen V, Wagner F, Nurnberg H (1986) The distribution of acid deposition on Germany. Experientia (Basel) 42:330-339

Vance YM (1983) Welcome home. Conservationist 44 N 2:18-21

Vangilder LD, Peterie TJ (1980) South Louisiana crude oil and DDE in the diet of mallard hens: effects on reproduction and duckling survival. Bull Environ Contam Toxicol 25:23-28

Vavilov NI (1937) The ways of Soviet science of selection. Izvestia of Acad Sci USSR, Biol Ser 3:635-670

Veprintsev BN, Rott NN (1979) Concerning genetic resources of animal species. Nature 280:633-634

Vernadsky VI (1965) Chemical structure of the Earth's biosphere and its surroundings. Nauka Press, Moscow, p 374

Vernadsky VI (1967) Biosphere. Mysl publishers. Moscow, p 376

Vernadsky VI (1977) Thoughts of the naturalist. Scientific thought as a planetary phenomenon. Nauka, Moscow, p 191

Vietmeyer N (1981) Man's new best friends. Quest 1:43-49

Vignes S (1981) Pollution radioactive and cancers: Risques aux faibles doses et normes de radio-protection. Med Biol Environ 9, N 2:22-27

Villiard P (1972) Wild mammals as pets. Garden City, New York, Doubleday, p 159. Publishers' Weekly 201, N 3:70

De Waal KJA, Van Den Brink WJ (eds) (1987) Environmental technology. Marinus Nijhoff. Dordrecht, Boston, Lancaster 826 pp

Wallis C (1981) Love among the condors. Time, 21.09.81:70

Walter W (1979) Vom Aussterben bedroht: die Geier Europas. Umsch Wiss Tech 79:750-751

Wassermann M, Ron M, Barcovici B, Wassermann D, Cucos S, Pines A (1982) Premature delivery and organochlorine compounds: polychlorinated biphenyls and some organochlorine insecticides. Environ Res 28:106-112

Watkins J, Klassen C (1986) Xenobiotic biotransformation in livestock: comparison to other species commonly used in toxicity testing. J Anim Sci 63:933-942

Watt KEF (1968) Ecology and resource management. A quantitative approach. McGraw-Hill, New York, p 463

Wayre P (1975) Conservation of eagle owls and other raptors through captive breeding and return to the wild. Breed Endangered Spec Captivity. London ea, pp 125-131

WCS (1978) Second draft of a world conservation strategy. IUCN-UNEP-WWF, Morges, pp 1-96

WCS (1980) World Conservation Strategy. Living resources conservation for sustainable development. IUCN, UNEP, WWF, FAO, UNESCO, p 48 + 5 maps

Weatherby AH, Cogger BMJ (1977) Fish culture: problems and prospects. Science 197:427

Webster B (1980) Songbirds decline in America. The New York Times, 12 August, pp C4-C5

Wehle D, Coleman C (1983) Plastics at sea. Nat Hist 92 N 2:20, 22-24, 26

Weinberger P, Greenhalgh R (1985) The sorptive capacity of an aquatic macrophyte for the pesticide aminocarb. J Environ Sci Health Part B Pestic Food Contam Agric Wastes 20:263-273

Weir JS (1971) The effect of creating additional water supplies in a Central African National Park. In: Sci Manag Anim Plant Communit Conserv. Oxford Univ Press, pp 367–385

Weisburd S (1985) Greenhouse gases en masse rival CO_2. Sci News 127:308

Weish P, Gruber E (1979) Radioaktivität und Umwelt. Fischer, Stuttgart, p 188

Wells S, Ryle R, Collins NM (1983) The IUCN invertebrate Red Data Book. IUCN, Gland, p 632

Werff M, Pryut JM (1982) Long-term effects of heavy metals on aquatic plants. Chemosphere 11:727–739

Westing AH (1982) Environmental quality. The effect of military preparations. Environment 24:2–3, 39

Wexler P (1988) Information resources in toxicology. Elsevier, New York, p 510

Whitaker JO, Schlueter RA, Tieben GA (1970) Effects of heated water on fish and invertebrates of white river at Petersburg, Indiana. Ind Univ Water Resour Res Cent Repr Invest, N 8, p 198

Whitcomb RF (1977) Island biogeography and "habitat islands" of eastern forest. Am Birds 31:3–5

Whitmore TC (1975) Tropical rainforest of the Far East. Oxford Univ Press, p 282

Whitmore TC (1980) The conservation of tropical rain forest. In: Soule M, Wilcox B (eds) Conservation biology. Sinauer, Sunderland, pp 303–318

Whittaker RH (1975) Communities and ecosystems. Macmillan, New York

Wilcox BA (1980) Insular ecology and conservation. In: Soule ME, Wilcox BA (eds) Conservation biology. Sinauer, Sunderland, pp 95–118

Wildemauwe C, Lontie JF, Schoofs L, Larebeke N van (1983) The mutagenicity in procaryotes of insecticides, and nematicides. Res Rev 89:129–178

Williams GM (1985) Identification of genotoxic and epigenetic carcinogens in liver culture systems. Regul Toxicol Pharmacol 5:132–144

Williams C, Harrison R (1984) Cadmium in the atmosphere. Experientia (Basel) 40:29–36

Wilson EO (1986) The value of systematics. Science 231:1057

Wilson EO, Willis EO (1975) Applied biogeography. In: Cody ML, Diamond JM (eds) Ecology and evolution of communities. Belknap Press of Harvard University, Cambridge, Mass, pp 522–543

Winkler M, DeWitt C (1985) Environmental impacts of peat mining in the United States. Environ Conserv 12:317–327

Winner W, Mooney H, Goldstein R (eds) (1985) Sulfur dioxide and vegetation. Stanford University Press, Stanford, p 593

Woo YT (1983) Carcinogenicity, mutagenicity and teratogenicity of carbamates, thiocarbamates and related compounds: an overview of structure-activity relationships and environmental concerns. J Environ Sci Health C 1:97–113

Wood JM (1982) Chlorinated hydrocarbons: oxidation in the biosphere. Environ Sci Technol 16:A291–A297

Wunz GA, Hayden AH (1973) Turkey renaissance. Nat Hist 82:86–93

Wurster CF (1968) DDT reduces photosynthesis by marine plankton. Science 159:1474–1475

Woodwell G, Hobbie J, Houghton R, Melillo J, Moore B, Peterson B, Shaver G (1983) Global deforestation: contribution to atmospheric carbon dioxide. Science 222:1081–1086

Workshop on the fate and impact of marine debris (1985) 27–29 November 1984. Honolulu. Hawaii, Executive Summary, p 1–11

Yablokov AV (1985) Problems of conservation of mammals in the USSR. In: Sokolov VE, Kuchuruk VV (eds) Teriology in the USSR. Nauka, Moscow, pp 180–197

Yablokov AV, Ostroumov SA (1983) Conservation of living nature: problems and prospects. Lesnaya Promyshlennost, Moscow, p 271

Yablokov AV, Ostroumov SA (1985) Levels of problems of conservation of living nature. Nauka, Moscow, p 175

Yablokov AV, Baranov AS, Rozanov AS (1980) Population structure, geographic variation and microphylogenesis of the sand lizard (Lacerta agilis). Evol Biol 12:91–127

Zach R, Mayoh KR (1982) Breeding biology of tree swallows and house wrens in a gradient of gamma radiation. Ecology 63:1720–1728

Zaletaev VS, Kostyukovsky VI, Novikova NM (1985) Ecosystems of Middle Asia under conditions of water regime reconstruction and problems of management. In: Problems of man-made effects on the environment. Nauka, Moscow, pp 62–67

Zehnder A, Zinder S (1980) The sulfur cycle. In: Hutzinger O (ed) The handbook of environmental chemistry. Springer, Berlin Heidelberg New York, 1A:105–146

Ziegler W (1982) Ionenkanale in planeren bimolekularen Lipidmembranen erzeugt durch das Herbizid SENCOR 70 WP. Biologia (Bratisl) 37:1071–1077

Zimmerman DR (1976) Endangered bird species: habitat manipulation methods. Science 192:876–878

Zimmermann HP, Doenges K, Röderer G (1985) Interaction of triethyl lead chloride with microtubules in vitro and in mammalian cell. Exp Cell Res 156:140–152

Zischke JA, Arthur JW, Nordlie KJ, Hermanutz RO, Standen DA, Henry TP (1983) Acidification effects on macroinvertebrates and fathead minnows (*Pimephales promelas*) in outdoor experimental channels. Water Res 17:47–63

Index of Authors

Subject Index